# The Origins of World War I
# *1871 - 1914*

Joachim Remak
University of California at Santa Barbara

COPYRIGHT © 2001 by Wadsworth Group. Wadsworth is an imprint of the Wadsworth Group, a division of Thomson Learning Inc. Thomson Learning™ is a trademark used herein under license.

Printed in the United States of America

**Wadsworth/Thomson Learning**
**10 Davis Drive**
**Belmont, CA 94002-3098**
**USA**

For information about our products, contact us:
**Thomson Learning Academic Resource Center**
**1-800-423-0563**
http://www.wadsworth.com

**International Headquarters**
Thomson Learning
International Division
290 Harbor Drive, 2nd Floor
Stamford, CT 06902-7477
USA

**UK/Europe/Middle East/South Africa**
Thomson Learning
Berkshire House
168-173 High Holborn
London WCIV 7AA

**Asia**
Thomson Learning
60 Albert Street, #15-01
Albert Complex
Singapore 189969

**Canada**
Nelson Thomson Learning
1120 Birchmount Road
Toronto, Ontario MIK 5G4
Canada
United Kingdom

ALL RIGHTS RESERVED. No part of this work covered by the copyright hereon may be reproduced or used in any form or by any means—graphic, electronic, or mechanical, including photocopying, recording, taping, Web distribution, or information storage and retrieval systems—without the written permission of the publisher.

ISBN 0-15-501438-2

The Adaptable Courseware Program consists of products and additions to existing Wadsworth Group products that are produced from camera-ready copy. Peer review, class testing, and accuracy are primarily the responsibility of the author(s).

# The Origins of World War I 1871-1914

SECOND EDITION

Der deutsche und der englische Friedensengel 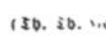

*A German View: A cartoon by Thomas Theodor Heine from the Munich satirical weekly* Simplicissimus *for February 8, 1909, showing Edward VII and William II wearing, as was the custom at such high-level meetings, the uniform of the other nation's army. The heading reads "The German and the British Angel of Peace," and the caption has William's angel saying "I would so much like to shake your hand. But would you mind taking off your boxing gloves first?"*

## Books by Joachim Remak

*Sarajevo, The Story of a Political Murder*

*The Gentle Critic: Theodor Fontane and German Politics*

*The Nazi Years, A Documentary History*

*The First World War: Causes, Conduct, Consequences*

*The Origins of the Second World War*

*Another Germany: A Reconsideration of the Imperial Era*
(with Jack Dukes)

*A Very Civil War: The Swiss Sonderbund War of 1847*

*O the gondolets, the mandolins, the twangling of the lutes,*
*The girls all dressed in crinoline among the flowers and fruits—*

*The flowers all symbolical, the lily and the rose,*
*And how the sherry blossomed on the end of grandma's nose.*

*The maiden sighs and turns away, the maiden she relents,*
*Attracted by the glitter of a pile of five per cents.*

*They danced beneath the arbors, they strolled upon the grass,*
*O never aware, O never aware of what would come to pass.*

THEODORE ROETHKE, *The Victorians*

# Preface to the First Edition

Few events compare, in impact and terror, with World War I. It destroyed two empires, that of the Habsburgs and that of the Ottoman Sultans. It altered two others, that of the Hohenzollerns and that of the Romanovs, beyond recognition. It created a host of new nations in Europe and overseas. It gave birth to Communist rule in Russia and provided the background for Fascism in Italy and elsewhere. In its course, roughly twice as many people were killed as in all the wars of the preceding two centuries added together, the Napoleonic Wars and the American Civil War included. It also maimed, in body or spirit, many of the survivors and laid the ground for an even more destructive World War II.

While the costs and consequences of World War I are matters of fairly general agreement, its origins still are not. In the immense literature on the causes of the war, a multitude of explanations, many of them sharply at variance with each other, have been advanced. But in this literature, for all the arguments and the contradictions, four major topics, or groups of problems, stand out. This book will deal with each of them in turn. The first is the conduct of European diplomacy between 1871 and 1914. The second involves the problem of certain basic underlying sources of danger, such as the alliance system, imperialism, or the role of public opinion. The third is the event that provided the war's immediate cause, the assassination of Archduke Franz Ferdinand at Sarajevo. The fourth concerns the conduct of European diplomacy in the five summer weeks that passed between Sarajevo and the opening shot of the war.

In the writing of this book, several debts have been incurred, which are gladly noted here. A number of publishers have kindly agreed to the use of copyrighted material. They are listed in detail in the bibliographical essay which follows the last chapter, where some of the obligations owed to other historians who have written on the war's origins, the giants in particular—Fay, Schmitt, Sontag, Albertini—are indicated too. Various friends, colleagues, and students have contributed facts and ideas, notably Donald Balmer, John Crampton, and Nosratollah Rassekh at Lewis and Clark College; D. F. O'Brien at the Menlo Academy, F. Smith Fussner at Reed College; Charles Jalavich at Indiana University; Wayne Vucinich at Stanford University; Alexander Callow,

Albert Lindemann, and Geoffrey Smith at the University of California at Santa Barbara; Peter Novick at the University of Chicago; Stephen Fischer-Galati at the University of Colorado; Donald Reid at Wesleyan University; and Frederick A. Breier at the University of San Francisco. Above all, however, the author's gratitude goes to Norman Rich at Michigan State University for his criticisms and suggestions, hard in substance, kind in manner, that were based on his reading of the manuscript.

J. R.

*University of California*
*Santa Barbara, California*
*April 1967*

# Preface to the Second Edition

New thoughts, new interpretations, and some new facts have led to the presentation of this revised edition. What has remained unchanged, however, even with the benefit of our added perspective, is the key importance of the First World War in revolutionizing European history and society.

*Joachim Remak*
*May 1994*

## CONTENTS

## Maps



## Illustrations

# CHAPTER 1

# The Age of Bismarck: European Diplomacy 1871–1890

## The Peace of Frankfurt and Some Consequences

"The origins," wrote the German author and filmmaker Alexander Kluge in his description of the battle of Stalingrad, "go back 30 days or 300 years."

Fortunately, the origins of the First World War can be pinpointed with essentially greater accuracy. More than anything else, what set in motion the course of events that led to disaster in 1914 was the Franco-Prussian War of 1870–71, and the peace that followed it. For it was that which opened a new age in European history. It meant the end of Napoleon III, and the re-establishment of a French republic. It also meant the achievement of German unity under the leadership of a triumphant Prussia. France, by the terms of the Peace of Frankfurt that formally ended the war in May, 1871, agreed to pay an indemnity of five billion francs to the new German Empire. France also agreed to surrender the province of Alsace, and a substantial portion of Lorraine. And France was powerless to prevent what it had been able to obstruct for centuries past—the creation of a unified and strong German nation.

It was a diplomatic revolution, but a revolution, there was reason to hope at first, that would not be allowed to escape from the hands of its principal creator—Otto von Bismarck, Prime Minister of Prussia since 1862 and Chancellor of the German Empire during the first two decades of its existence from 1871 to 1890. Bismarck's aim, after Frankfurt, was peace. The Germans, he felt, had what they had set out to achieve, and what had so long eluded them, most notably in 1848—union and power. Germany henceforth should consolidate her gains and seek no new ones, not in Europe and not overseas. Her strength and fame were alarming enough to other nations as things stood. What Bismarck now wanted was calm; the new equilibrium of Europe ought not to be disturbed again. Bismarck's motives were not necessarily those of a philosophical opponent of force. They were, rather, the motives of a supremely astute politician who believed, rightly so, that the interests of Prussia-Germany, under the circumstances, happened to be best served by a long period of peace.

No matter what the motives were, the effects were largely happy ones. They would have been happier still, had not two problems remained that not all the skills of Bismarck could resolve. One was the

*Otto von Bismarck at the age of 50. Watercolor sketch by Adolph von Menzel, 1865. (Kupferstichkabinett der Staatlichen Museen zu Berlin.)*

problem of Alsace-Lorraine, the other that of the change in Europe's balance of power.

After Prussia's victory over Austria in 1866, Bismarck, in a scene heavy with emotion and barely suppressed violence, had fiercely argued against the victor's peace which the king and his generals had in mind. When the meeting was over, he collapsed with a crying spell; a while later, he contemplated suicide. But in the end, he had his way. The peace with Austria was a generous one, and the absence of any vindictive conditions left the future clear, as Bismarck had intended, for friendship with the former enemy.

There was no repetition of the scene after the Franco-Prussian War. The pressures for a Prussian annexation of Alsace-Lorraine were too strong, the grounds looked too valid. Prussia's king did not wish to be robbed of his triumph again. The generals argued that they needed the fortress of Metz in Lorraine to secure Germany against possible French attack. Public opinion demanded some concrete and, it seemed, entirely legitimate, fruits of victory; had not the name of the Alsatian capital more often been spelled Strassburg than Strasbourg?

> Germany's river, Strassburg's *Dom*
> Return them to their German home

wrote Heinrich von Treitschke, teaching history and patriotism in Berlin.

Faced with this chorus of king, generals, professors, and ordinary voters (and in his heart of hearts sharing their sentiments himself), Bismarck acquiesced. France, he consoled himself, would not love Germany in any event, no matter what the terms of the peace were. It was a specious argument, and the Chancellor was aware of it. For, plausible though the reasons advanced for a victor's peace might seem, all of them, in the end, proved to be very bad indeed.

Bismarck's moral capital was immense; some plain words on his part might have altered public opinion and the king's as well. The Alsace might be German in language, but in sentiment it was not. Kellermann, Kléber, Ney—the generals of the Revolution and of Napoleon, those "bravest of the brave"—had opted for France just as Steuben and the Muhlenbergs had opted for America. 1789 meant more than 1870; Paris was closer than Berlin. And what held true for Alsace trebly held for Lorraine. Before long, Treitschke was to become very angry with Germany's new citizens. The Alsatians' desires were "misguided"; they should be made to realize that they were Germans, if need be by force. But what of military necessity? Metz, said the chief of the Prussian general staff, would be worth an army of 120,000 men. In a sense, it was a brilliant estimate of the situation; after Frankfurt, Germany possessed an ideal military frontier in the west. In 1914, a very small German force would be able to hold back a major French army there. In another sense, it was a catastrophic misjudgment, since the establishment of precisely this frontier would be one of the main reasons why the two armies were advancing on each other in 1914. Too many Frenchmen would not resign themselves to the loss of Alsace-Lorraine: "Always think of it, never speak of it." In fact, speak of it; shroud in black the statue of Strasbourg among the cities of France represented on the Place de la Concorde; shroud her in black until 1871 might be undone.

Here, then, was one of the basic sources of discontent after Frankfurt. The other was that Europe's balance of power had materially changed. Of the five powers that had been great before 1871, only England and Russia retained roughly their former places. Austria, once formidable, had become relatively weak, and so had France. Conversely, Prussia—or rather Prussia-Germany now—had risen to the rank of a major power. So, in some ways, had a sixth—Italy, whose unification was complete now that her leaders had taken advantage of the diversion of the Franco-Prussian War to occupy Rome. Mainly, however, it was Germany's new position that was of concern. Italy was a power largely by courtesy, power in the nineteenth century as in others being mainly measured in guns rather than in Giottos, that is, by the ability to put an effective army into the field in case of need.

Bismarck, being aware of both problems, took the approach that they were not insoluble. Nations can ultimately reconcile themselves to territorial losses. Shifts in the balance of power do not automatically result in new wars. His answer, over the period of his chancellorship, was a threefold one. As for France: try to conciliate her. Aid French ambitions elsewhere; perhaps Alsace could be forgotten in time. "I want you to forgive Sedan, as after 1815 you came to forgive Waterloo," he said to the French Ambassador. To this end, Bismarck was to encourage French expansion into North Africa and China, supporting the French position in Tunis in the 1880s even at the risk of antagonizing Italy, Germany's incipient ally.

As for the change in the balance of power: let the new Germany speak and act softly; assure the world of her pacific intentions. Thus Bismarck was to hold out as long as he could against the acquisition of colonies. When one of many visitors came to see him to urge that Germany, too, go exploring for ivory and ebony, Bismarck heard him out, and then said: "My map of Africa lies in Europe. Here is Russia, and there is France, and here in the middle are we. That is my map of Africa."

As for both the problem of France and of Europe's changed power structure: see to it that any French desire for revenge be kept in check by France's inability to find allies. Isolate her diplomatically, unless the alliance be with Germany—a possibility that, for all its remoteness, Bismarck did not exclude. But for the moment, keep France as friendless as Germany should be gregarious. International politics, he told a fellow diplomat, mainly reduced themselves to a simple formula: "In a world governed by five powers, try to be *à trois*."

## The New Alliance System: The Three Emperors' League of 1873

The three Bismarck first joined together were Germany, Austria-Hungary, and Russia. A series of suitably festive state visits in 1872 involving Germany's William I, Austria's Francis Joseph, and Russia's Alexander II, followed by more prosaic and formal negotiations a year later, established the Three Emperors' League. Their governments would consult with one another on matters of mutual interest; and more importantly, they would maintain friendly neutrality—or even aid each other actively—in case any one of them should be attacked by another power.

The League made a fair amount of sense. There was Austria's need, particularly strong in view of her domestic problems, for a reconciliation with the Prussian enemy of 1866. There were the shades, particularly dear to the Russians, of the Holy Alliance, and tsarist hopes that the more conservative of the Big Five might once again make common cause against political innovation. There was, above all, Germany's desire to banish the specter of a two-front war. Should war ever come again, France, by herself would not be a mortal threat; the combination of Russia and France would be. However, where the Three Emperors' League made less sense was in that of the three partners pledged to friendship, two, Russia and Austria, were in truth very much getting in each other's way in Southeastern Europe. Thus the League's first real test, the Congress of Berlin, nearly turned out to be its last.

## The Decline of the Ottoman Empire, The Russo-Turkish War, and the Congress of Berlin of 1878

The events that led up to the Congress were of an order to which Kluge's "30 days or 300 years" did fairly apply, though with a preference for the latter, for the roots here indeed went deep. The Turkish Empire, which had once reached from Hungary in the West to Persia in the East, from the Crimea in the North to Egypt in the South, had for some time been in a state of decline. The troubles that plagued it were many: old age, a corrupt and inefficient sultanate, a multiracial, multinational structure in a period when the fashion was for national independence.

The empire, while shrinking—the Crimea long ago had been lost to the Romanovs, Hungary to the Habsburgs—still struggled to maintain itself in the Near East and in the Balkans. Its principal nineteenth-century antagonists were of two sizes. There were the small Balkan nations, or those that meant to become nations—Greece, Albania, Bulgaria, Montenegro, Serbia, Rumania—who were openly anxious to be free of Ottoman rule, or to add to what independent territory they already held. Then there were the giants: Russia, Austria-Hungary, and now and then Great Britain and France. Their approach was less direct. Publicly, they spoke with sorrow about the woes of the supposedly sick man of Europe. Privately, they tended to hope for a speedy deterioration of his condition so that, in the case of Austria and Russia, they might fall heir to his position in the Balkans, and in that of Britain, France, and of Russia again, to that in the Near East.

Turkey's main chance for survival lay in the safety offered by her ill-wishers' numbers. Had there been fewer presumptive heirs, the empire might well have been destroyed long before it actually was. As matters stood, Turkey possessed a certain amount of insurance in the inability of her adversaries to agree on a division of the spoils. Besides, when all was said and done, and this tended to be the British position in particular, enabling a weak Ottoman Empire to survive was preferable to allowing one of the rival powers of Europe to increase its strength at the Sultan's expense.

It was a precarious survival, however. More than once, Turkey had to put up with the intervention of the European powers in her internal affairs. The Greek War for Independence in the 1820s and the Crimean

War three decades later provided some familiar examples. The 1870s now were to see one more.

In 1875, a revolt against Turkish rule broke out in the twin provinces of Bosnia and Herzegovina. This time, the usual pattern of unrest, rebellion, bloody suppression, and relative peace was broken. First, Serbia and Montenegro took up arms in defense of their Slavic brothers in Bosnia. Next, the revolt spread to Bulgaria, which had recently been granted some degree of autonomy within the Turkish Empire. The reasons for rebellion were much the same here as in Bosnia: economic discontent, a strong feeling of national consciousness, and hatred by the largely Christian population for the Moslem Turks. Finally, Russia decided, early in 1877, to join the battle against the Turks.

Russia's motives were a mixture of the old and the new. It had long been a desire of the Russians to see a weakened Ottoman Empire. What Russia had needed and wanted for centuries were the Dardanelles. The Russians could see no other way of having an assured access to the Mediterranean and thus to the world's commerce in times of peace, or of deploying their fleet with any true degree of effectiveness in times of war. "Would you deny me the keys to my house?" Alexander had asked Napoleon when they had met at Tilsit in 1807 to divide the world. Napoleon would, and so would all the other statesmen of Europe. The aim remained.

The new but hardly contradictory motive was Pan-Slavism. It was Russia's historic mission, or so went the Pan-Slav argument, to protect and ultimately liberate the Slavs of Southeastern Europe presently living under foreign rule, Austrian or Turkish. All Slavs were brothers, and the term Slav was interpreted generously to include Poles, Bulgarians, Serbs, Slovenes, Czechs, Croats, Slovaks. Nor did all Pan-Slavs stop even there: some envisioned a truly Grand Federation of the East, to include Greece (Orthodox in faith as was Russia), and above all, Constantinople. Constantinople would then receive its rightful name of Tsarigrad, the Imperial City, and union would reign between Moscow and Byzantium, between the Third Rome and the Second.

At times, the Russian government would frown on Pan-Slavism. Its program, obviously, was about as harmless as a forest fire. At other times, the government would smile on it. Pan-Slavism provided a useful diversion from tsardom's internal difficulties; it also had its uses in the pursuit of some very concrete Russian ambitions abroad.

During the early stages of the Bosnian and Bulgarian troubles, Alexander II had restrained the Pan-Slavs, but ultimately, the pressure

The Key to My House, ca. 1914. *The Dardanelles: The strait connecting the Sea of Marmara with the Aegean—a passage long considered vital by Russia for access to the Mediterranean and thus to the world's trade.*

for war proved too strong. The Pan-Slavs certainly had been right in one regard. The Turks were tired; the war was not a difficult one to win. Quite possibly, it was, as a foreign military observer put it, a battle between the one-eyed and the blind, but in a contest of this sort it is good to have one eye. A surprising Turkish stand at the Fortress of Plevna did slow the Russian advance and kept Russia from overrunning all of European Turkey, but even so, the Turks had little choice but to ask for peace and to accept it on Russia's terms.

The Treaty of San Stefano, which the two nations signed in March 1878, reflected the extent of Russia's triumph. It provided for autonomy or outright independence for the Ottoman Empire's Balkan holdings—for Bulgaria, Rumania, Serbia, and Bosnia-Herzegovina—and rewarded Russia both with territorial annexation (most notably of Bessarabia in modern Rumania) and with a vast indemnity. Turkey had effectively been expelled from Europe, its influence replaced by that of Russia. To achieve that goal, the Russians had acted with an astonishing disregard for anyone else's interests. Europe, quite predictably, took alarm.

The powers most immediately concerned were Great Britain and Austria-Hungary. England's role was the slightly more ambivalent of the two. Much of British public opinion tended to have a good deal of sympathy for the nations rising in rebellion against Turkey. William Gladstone, four times prime minister in the last quarter of the nineteenth century, had published a celebrated pamphlet in 1876 on the "Bulgarian Horrors." In it, he damned the Turks for their cruelty and demanded what the Russians were in fact to force the Turks into granting: autonomy for the Balkan Christians. But Gladstone, Liberal and humanitarian, was not in office at the time. Benjamin Disraeli, Conservative and empire-minded, was; reasons of state mattered more to him than philanthropic impulses. It should be Britain's principal object, he wrote, "to keep the Russians out of Turkey, not to create an ideal existence for Turkish Xtians." Great Britain's interests in the Near East, Great Britain's concept of the balance of power, both precluded the acceptance of Russia's gains. Therefore, just prior to San Stefano, Disraeli had dispatched units of the British fleet to Constantinople; thus, he now was willing to go beyond naval demonstrations and contemplate war with Russia. The Balkan Christians might as well cease complaining; "autonomy for Ireland would be less absurd."

British opinion, at the spectacle of Russia's victories, came to side with Disraeli. The poor Bulgarians were forgotten; the Turks promoted from bloody oppressors of defenseless Christians to gallant fighters against the brutal Russians, England's ancient adversaries. It was the crisis that enriched the language with the word "jingoism."

> We don't want to fight,
> But by jingo if we do,
> We've got the men,
> We've got the ships,
> We've got the money too!

wrote the otherwise obscure British poet G. W. Hunt on the occasion.

Were Britain to fight, at least one continental power might join her, Austria-Hungary. Austria's motives for opposing Russia's triumph were less ambitious than were Great Britain's. Vienna mainly wished to hold on to its possessions rather than add to the splendors of the empire. Austria's basic problems were not all that dissimilar from Turkey's. It was a multiracial state, comprising within its borders at least a dozen different major nationalities and cultures, and nearly all of the world's religions. Economically and historically, the empire made a very great deal of sense. Beside, it could be said of Austria what also (although with more reservations) could be said of Turkey: any institution that has lasted for some centuries and met its needs with some success deserves respect. Still, many of the nationalities that made up the empire took a less Burkean view of things. In varying degrees, they were feeling dissatisfied with their role in the monarchy, resenting what they considered to be excessive German and Magyar domination. In the case of the Magyars, these grievances could be only too well documented: Budapest's nationality policy was distinguished neither by its generosity nor by its tolerance. Vienna's record was far better, as good as that of any pluralistic empire, Rome included. But the restiveness remained, particularly among the monarchy's Slavic population. The men in charge of Austrian policy, in consequence, were aghast at San Stefano. What were they to make of Russia's broken word? Bosnia and Herzegovina, the tsar had promised before he was sure of victory, were to go to Austria. How was Austria-Hungary to live with Russia's suddenly overwhelming presence in the Balkans?

The Bosnian question was at the core of it. Again, it was not so much that Austria coveted new territory. The monarchy was having enough to worry about with what it already encompassed. It was rather that Vienna could visualize what would happen if the newly autonomous region should one day join Serbia. The greater Serbian kingdom thus created would be a frighteningly attractive magnet to the empire's South

Slav citizens. Perhaps Austria, too, should give battle to undo San Stefano—without any bellicose lyrics, sadly rather, as became the Austrian mood—but battle nonetheless.

The man who in the end was instrumental in keeping anyone from fighting was Bismarck. Peace was his platform; no Balkan troubles would dislodge him from it. In the summer of 1878, the powers agreed to meet at Berlin, to reconsider, under Bismarck's chairmanship, the provisions of San Stefano. The Congress, to the shock of the Russians, all but demolished the treaty. Not that everything went wrong for the Russians at the Congress of Berlin. The independence of Serbia and Montenegro was confirmed, although their borders were slightly redrawn. Confirmed, too, were most of Russia's direct gains, notably Kars, Batum, and Bessarabia. However, the principality of Bulgaria was deprived of its access to the Aegean and reduced to one-third the size envisaged at San Stefano. Moreover, Russia's gains were balanced by compensating the other powers at Turkey's expense. Britain received Cyprus. The French received clearance to move from Algeria into Tunis. The Austrians received Bosnia and Herzegovina, nominally, to "occupy and administer"; actually, there was no need to spell out how many tenths of the law comprised possession.

Only the Germans demanded, and the Italians received, nothing. Germany was satisfied to play the role of "honest broker"; here was as good an opportunity as any to display the new Reich's self-restraint. As for the other newly unified nation, enforced moderation seemed the essence of wisdom. The Italians, said Bismarck, had such a large appetite and such poor teeth.

True unhappiness over the results of the Congress was reserved, however, to the Russians and to the Balkan nations. The Russians felt deserted by the Germans at Berlin—where was the support they should be receiving from their ally of the Three Emperors' League? Equally bitter was the sentiment in the Balkans. For the most part, independence or autonomy stood confirmed, but the new borders seemed to have been drawn with considerable disregard for ethnic realities and national ambitions. The Rumanians resented the loss of Bessarabia. The Greeks felt that their aspirations had not received a proper hearing. The Serbians mourned for the loss of Bosnia and Herzegovina. The Bulgarians would not reconcile themselves to the disappearance of the mirage of a Greater Bulgaria conjured up at San Stefano.

It was easy enough to counter these objections. Russia, after all, had brought her Berlin defeat upon herself by her greed at San Stefano.

The powers of Europe had valid cause for being more interested in preventing the diplomatic revolution that would have resulted from the Russian-designed destruction of European Turkey than in establishing ethnic justice in Southeastern Europe. Besides, not all the wisdom of Solomon could have defined the meaning of justice in this case; the person did not exist who could have drawn indisputably equitable borders in this part of the world. Greece, Rumania, and Serbia certainly had not been noticeably happier with San Stefano than they were with Berlin.

All such arguments, however, did not change the fact that taken together, the Balkan Wars, San Stefano, and Berlin provided two of World War I's long-range causes. One was that the unresolved discontents of Southeastern Europe were to be at the root of several subsequent Balkan crises, and then of Sarajevo. The other was the ominous estrangement between Russia and Germany.

## The Dual Alliance of 1879

Publicly, Bismarck had been charm itself at the Congress of Berlin. Privately, he was seething. For three years now, ever since the beginning of the Balkan imbroglio, he had been doing his utmost to prevent a break between Austria and Russia. He had no wish to take sides between his allies. Never cut the wire to St. Petersburg, was his maxim. Never cut the wire to Vienna either.

Berlin had ended that. He had labored hard at the Congress to find a fair compromise. He had labored in vain. The Three Emperors' League was dead. The Russians, ready as always to indulge their sense of grievance, were fuming at him, from tsar on down to inspired press.

Forced to make the next move now, Bismarck replaced the Three Emperors' League with a bipartite treaty with Austria, the Dual Alliance of 1879. Designed to last for an initial period of five years, it was renewed periodically, and thus did not expire until the two imperial regimes signing it did, in 1918. It was a defensive alliance under whose terms the signatories agreed to come to each other's aid, should Russia attack either country. Should a country other than Russia be the aggressor, the treaty partner would maintain a position of benevolent neutrality, unless Russia should join the war, in which case armed assistance again was mandatory.

The Dual Alliance has received strong criticism. Bismarck, his critics charged, had tied the fate of a vigorous Germany to that of the troubled Habsburg monarchy. He also, by opting for Austria, was forcing the Russians to seek an alliance elsewhere, specifically with France.

Bismarck took both of these objections into account. As for tying Germany's fate to that of Austria, the Germans would plainly have to be on guard here. But who, after all, was really tied to whom? Every alliance, said Bismarck, had a horse and a rider, and in this one, Germany should easily be in the saddle. What the rider must do was remain alert and restrain the horse. Germany ought never to encourage, he wrote the German Ambassador at Vienna in a typical instruction of the mid-eighties, "the temptations, to which the Austrians are prone, to exploit the situation in order to use the armed forces of Germany for the benefit of Hungarian and Catholic ambitions in the Balkans." The term "defensive" in the alliance would have to mean just that. If Austria should ever be a victim of Russian attack, German aid would freely be given, "but an *attack* on Russia is something we will not engage in, nor will we, in case Austria should undertake it, consider the alliance to be operative. . . . For us, Balkan questions can never be a motive for war."

But what of the offense that the special Hohenzollern-Habsburg partnership might give the Russians? Here, too, Bismarck was able to think several moves ahead. He did not believe that Russia could or would turn to the West. Ideologically and diplomatically, too much divided the land of the tsars from Britain and France. He did believe that Russia, when faced with a firm commitment between Austria and Germany, and thus with the prospect of diplomatic isolation, would trade resentment for realism. In diplomacy as in politics, there can be no permanent enmities.

## The Three Emperors' League Renewed, 1881

His calculation proved to be correct. The Dual Alliance had hardly been signed when the Russians began to suggest a Russo-German alliance. Negotiations were slow, for Bismarck wished Austria to be included. He was to have his way. In 1881, Russia, Germany, and Austria formally signed an agreement reviving the Three Emperors' League. The treaty's scope was broader than that of its predecessor. The contracting parties agreed, with no defensive clause this time, that if any one of them should "find itself at war" with a fourth power, the other two would remain neutral. This went far to relieve Germany's dread of a two-front war against both Russia and France. It also helped to alleviate Russia's fear of an Austro-British military coalition against her that had been such a distinct possibility in 1878. In one respect, however, the new Three Emperors' League was narrower in scope. It contained a major qualification concerning the applicability of its provisions in that region of Europe where Austro-Russian interests were most at variance. The promise of neutrality, read the treaty, was to apply in case of war against Turkey only on condition that the three signatories first agreed on "the results of this war" or, more plainly put, on a division of the spoils.

One Turkish spoil already distributed in 1881 was Bosnia-Herzegovina. Austria, under the terms of the treaty, received the right to annex the two provinces "at whatever moment she shall deem opportune." Russia, in return, received assurances that the Straits were to remain closed to warships of all nations; Russia was to be secure from naval attacks from the Black Sea. This was a matter of particular concern to the Russians, who had done little to rebuild the Black Sea fleet destroyed in the Crimean War and thus had good reason to fear a British fleet sailing through the Straits, guns pointed at Odessa or Sevastapol.

It was a good treaty in all. As seen from Russia and Austria, some very specific gains could be noted—the Bosnian concession and the Black Sea clause, and beyond that, the implicit establishment of an armistice in the Balkans. Obviously, much would depend on the spirit in which the treaty would be observed, but chances were fair that the Balkan truce might turn into genuine peace. As seen from Bismarck's chancellery, the gains were less direct but even more pleasing: Germany was *à trois* again, France *tout seul*.

## The Triple Alliance of 1882

In fact, the Germans were to gain one more ally. In 1882, attracted by power and offended by France, Italy asked to join the Austro-German Dual Alliance. The Italians, eager for empire, had for some time had their eyes on Tunis, but the French hand was quicker than the Italian eye. In 1881, the French, moving from Algeria, established a Tunisian protectorate. The Italians were outraged, the more so since they were unable to offer any effective resistance to France. The consequence was a search for allies, even for so unlikely an associate as Austria, whose popularity was no greater than that of France, since the Trentino and Trieste—"unredeemed Italy"—were still in Habsburg hands.

Bismarck's enthusiasm was muted. He feared that once the Austrians made their peace with Rome and felt secure to the southwest, they might again become hard to handle in their Balkan conflicts with Russia.

But in the end, he agreed to a defensive alliance. Its most specific provisions dealt with France. Should France, "without direct provocation," attack Italy, Austria and Germany would come to Italy's aid. Should France attack Germany, Italy, in turn, would provide military support to her ally.

Of the five great European powers with which he had to reckon, Bismarck now was on good terms with one: Great Britain (no part of the Triple Alliance, stated a supplementary declaration, should "in any case be regarded as directed against England"), and formally allied with three others: Italy, Russia, and Austria. If only he could keep his friends from coming to blows with each other, he would have no cause to fear his enemy. But controlling all of his friends all of the time proved to be beyond even his powers. In 1885, a new Balkan crisis unfolded which once more forced Bismarck into the role he was least inclined to play, that of arbiter between Russia and Austria.

## The Bulgarian Crisis and the End of the Three Emperors' League, 1885–1887

The Balkan nation that had lost more than any other at the Congress of Berlin had been Bulgaria. In 1885, the Bulgarians saw an opportunity to recover much of what they had lost seven years earlier. A revolt had broken out in Eastern Rumelia, a region Bulgarian in population and sympathy. It had, accordingly, been made a part of the Greater Bulgaria created at San Stefano, but the Congress of Berlin, while granting it autonomy, had returned it to Turkish suzerainty. The Bulgarians, understandably, decided to support the new rebellion, which had the exchange of autonomy for union with Serbia as its aim.

Bulgaria faced two major obstacles. This time, there was no Russian aid; the new nation had shown itself far too independent in spirit and action for the Tsar's taste. There also was Serbia's hostility. Belgrade was not sure that it could live with a too-triumphant Bulgaria; it also sensed that this might be the time to settle the fate of Macedonia, that ancient and bitter issue among Serbia, Greece, and Bulgaria.

The Serbians decided on war. Peaceful habits would have become them more. With speed and skill, they were routed by the Bulgarians. Only Serbia's alliance with Austria, concluded in 1881, saved the country from utter disaster.

But Bulgaria's triumph was too great for Russia's comfort. The tsar now threatened to occupy Bulgaria and to put a more pliant ruler on the throne. This move brought Austria onto the scene again. As long as it had been a fight between Serbians and Bulgarians and the principal object was Eastern Rumelia, the Austrians could afford to be anti-Bulgarian. Now that Russian intervention threatened, they no longer could. Self-defense seemed to require a stand in favor of Bulgaria; Pan-Slavism had to be stopped somewhere.

The outcome was a diplomatic victory for the Austrians. Ably assisted by Great Britain—where the prospect of a Russian-dominated Balkans was no more attractive now than it had been at the time of Berlin—Austria prevented the Russian move. The Russians were infuriated, directing their anger at friend and foe alike. In the face of the many appeals that came to him from Vienna and St. Petersburg in the course of the crisis, Bismarck had tried hard to remain neutral. When that proved impossible, he had shown some mild sympathy for the Russian position. But mild was the word; for the fear of losing his Austrian ally kept him from giving the Russians the all-out support that they demanded as their right. In consequence, Russia now despised the Germans nearly as much as she did the Austrians, and in 1887, as the Bulgarian crisis was ending, so, too, was the Three Emperors' League.

## From the Reinsurance Treaty to the Fall of Bismarck, 1887–1890

The League had been signed for a period of as many years as there were emperors in it. In 1884, when it first came up for renewal, the Russians were happy to sign again. In 1887 they were not. But their respect for realities was strong enough to make them propose a bipartite agreement to Germany. It was not the kind of treaty Bismarck would have chosen, but it was better than none. Hence, in 1887, he signed what came to be known as the Reinsurance Treaty with Russia. If either power should find itself at war, the other would maintain a benevolent neutrality. The only exception was to be aggressive war. Should Russia attack Austria, or should Germany attack France, the treaty partner would have full freedom of action.

The Russians received what looked like a great deal for having insured the Germans against a two-front war. Germany recognized Russia's preponderant influence in Bulgaria; reaffirmed the principle of the closure of the Straits; and, better than that, promised moral and diplomatic support for any measures the tsar might take to control the key of his Empire.

Bismarck seemed to have given far more than he received. Seemed to have, for in truth Russia's gains were largely paper gains. In perhaps the most intricate series of diplomatic maneuvers of his career, Bismarck was seeing to that. Even before the formal completion of the Reinsurance Treaty, he had been party to bringing about a Mediterranean Agreement between Great Britain, Austria, and Italy, in which the three powers undertook to maintain the status quo in the Balkans, the Near East, and North Africa. In this way his promises to Russia were effectively neutralized, for what he had in truth granted to the Russians was something that he knew the British and Austrians could be relied on to withhold from them. Next, he allowed the terms of his various secret agreements to leak out. To impose some sensible caution on the Russians in the Balkans, he let them know about the Dual Alliance with Austria. So that they might have no doubt about the strength of the coalition that would face them should they try to pocket the house key, he let them know about the Mediterranean Agreement as well. The Austrians, in their turn, had the Reinsurance Treaty brought to their attention; it would be a good reminder that German aid was not automatic. Finally he let the French learn about both the Triple Alliance and the Mediterranean Agreement; it would do them no harm to realize that the world contained potential enemies other than the occupiers of Alsace.

While these maneuvers may have been several shades too complex and too subtle, and leavened with rather more than the ordinary amount of diplomatic duplicity, what needs to be said for them is that they worked, that they served Bismarck's overriding motive of maintaining peace. The multiplicity of alliances did make for caution; Vienna as well as St. Petersburg found it politic to restrain their ambitions. How well the system would work without its original designer was a different question. The answer would come all too soon. In 1890, Bismarck had to step down as chancellor. Ostensibly, policy differences between him and the new Emperor, William II, played a major part in his dismissal. Actually, the principal reason—and for once we should have some sympathy for William II—was that the young monarch found it understandably difficult to get along with a national monument for a chancellor. Still, the pilot was gone. Europe would not be the same again.

Bismarck's achievement, as we look back on the two decades of his chancellorship, is as astonishing as ever. He had voluntarily restrained Germany's quest for power. The sky was the fool's limit, not his. He had built an alliance system that left France without allies, and Germany with as many as any one power could manage. The diplomatic alignment

*Bismarck's departure from office in 1890 provoked this reaction in* Punch, *the London satirical weekly. It would become one of the most famous—and prophetic—cartoons of its time. The artist is John Tenniel, celebrated both as a long-time contributor to* Punch, *and the illustrator of* Alice in Wonderland. *(The melancholic onlooker is the new Emperor, William II.)*

of 1914 was in a future more remote than the quantum theory. Rus. was still allied to Germany, Austria was ready to fight in the defen of Serbia, and England and Russia invariably found themselves ( opposite sides. Of the conflicts that remained, some seemed perfect capable of pacific solution, and those that perhaps were not, still di not seem worth the cost of a general war. As much as any one mar Bismarck had helped to create the late nineteenth-century mood c Europe, which appeared to take it for granted that war was an aberration and that peace was man's natural estate.

# Chapter 2

# Triple Alliance and Triple Entente: European Diplomacy 1890–1914

L'Impudique Albion

## The Initiative Passes: The Reemergence of France

That Bismarck's successors would not be his equals might have been expected; that quite so much would divide them from him was another matter. William II by no means lacked intelligence, but he used it badly. (A trait that brought to mind the classic appraisal of a junior army officer by his superior as "a man of many talents who does not know what to do with them.") For all of the Kaiser's occasional charm and judgment, he was unstable, impulsive, incapable of sustained effort, uncertain of his aims, and given to bluster. His public speeches had a way of disconcerting his friends and delighting his enemies. They abounded with talk of might and "shining armor"; of a German future that lay on the water; of "days of splendor" toward which he was leading his nation, although along that road to glory there would then be more than one instance of "my and my people's exhausted patience." They contained supreme imbecilities such as his claims to divine right—he was "the instrument of the Lord," who would go his way "without regard for daily opinions and intentions"—or his famous exhortation to the German troops leaving for China after the Boxer rebellion:

> You must know, my men, that you are about to meet a crafty, well-armed, cruel foe! Meet him and beat him! Give no quarter! Take no prisoners! Let his life be forfeit if he falls into your hands! Just as, a thousand years ago, the Huns under their king Attila made a name for themselves which makes their power resound in legend and fable to this very day, so may you affirm the German name in China in such a manner that never again will a Chinese dare as much as look askance at a German!

Not all his speeches were this outrageous, nor did the spirit really match the bluster. But that hardly excused them. As a member of William's Court, and a sharp observer of the contemporary scene, the Baroness von Spitzemberg wrote in her diary at the time, "There is something dreadful about this frenzy of speechmaking and this boasting, especially when it involves things that the emperor never has any

*The Unlikely Allies? British antipathies to France were equaled by those of the French to Britain. There is, for instance, this cartoon by Jean Veber, prompted by published reports of Britain's running concentration camps for the defeated Boers in South Africa. It appeared in* L'Assiette au Beurre *in September 1911, and the caption reads, "Britain the Shameless." (The features are those of King Edward VII.)*

intention of *doing*." And what was so perilous about this "frenzy of speechmaking" was that there now was no Bismarck to counsel or restrain him. In part, this was the Emperor's own fault: flattery carried more weight with William II than talent. However, one need not exaggerate that point; the men who held office under him were not incompetents or fools, and while a French observer would call the man who followed Bismarck into the foreign ministry the *ministre étranger aux affaires*, this was moderately funny but untrue. He was by no means "foreign to affairs"; he as well as his colleagues knew their subject well enough; all they lacked was genius, and later, courage. And they were aware of this. "What kind of jackass will dare to be Bismarck's successor," General von Caprivi once asked a friend. When, not much later, William II asked Caprivi to succeed Bismarck as chancellor, the general's initial reaction was to reply that Bismarck dead and carried in front of the Emperor like the Cid would be of more use to the country than he, Caprivi, alive. Having accepted the office for all his misgivings, it was Caprivi who also coined the famous description of Bismarck as the man who alone could keep five glass balls in the air at the same time. There was an element of truth in the image, nor is it easy not to sympathize with Caprivi. The analogy of the five balls certainly turned into a comfortable cliché until a young Swiss historian, Leonhard von Muralt, suggested that if it were true, Caprivi had no right to remain in office. The number of balls, after all, had not decreased just because Caprivi said that no one but Bismarck could juggle them.

Another power was now willing to try some difficult juggling, the power that had been in apparent eclipse during the past two decades—France. What Germany most notably lacked in the quarter century before 1914, France possessed—diplomatic genius. No one person would dominate the scene in Paris as Bismarck had in Berlin, but among an abundance of talent, four men at least stand out: Delcassé, Poincaré, and Paul and Jules Cambon.

Théophile Delcassé, long a parliamentary deputy, served as colonial secretary in the early 1890s, as foreign minister from 1898 to 1905, then as secretary of the navy, and, after 1914, as foreign minister again. His admirers' epithet, "France's answer to Bismarck," was far from wrong. He could move among very different allies with ease. He could tell the essential from the ephemeral. He could make some clear choices where indecision, while pleasant, was folly; it was Delcassé who faced the fact that France could not afford the luxury of forever living at cross-purposes with both Britain and Germany and who, at the risk of antagonizing considerable parts of French public opinion, set to turning one of the two into a friend.

He was splendidly supported by the two Cambon brothers. Paul Cambon, during the period that Delcassé was foreign minister and for some years thereafter, was ambassador to Great Britain; Jules Cambon, after holding office as governor-general of Algeria, successively served as ambassador in Washington, Madrid, and Berlin. Their reports are models of lucidity. Their approach to diplomacy brings to mind Bismarck's more celebrated than heeded dictum about politics as the art of the possible; it was Paul Cambon who would say, apropos of his recommendation for reaching a *rapprochement* with Russia: "If you cannot have what you like, you must like what you have." He would have generous opportunity to work toward that end, for one other aspect of French foreign policy of the period that deserves special note was its continuity. Cabinets came and went, but foreign ministers and ambassadors remained. Delcassé served throughout the lives of six subsequent ministries; Paul Cambon would represent France at the Court of Saint James for twenty-two years.

Finally (or rather finally in this select list only) there was Raymond Poincaré, senator, minister of finance, premier, and, in 1913, President of the Republic. A native of Lorraine, Poincaré was as passionate a Frenchman and republican as Bismarck had been a Prussian and a monarchist. He was a man of shrewdness, probity, great energy, and that particularly French combination of rationalism and conservatism; some writers have seen him as the personification of all that has been best in bourgeois France. He was intent on building up French strength against Germany, not so much because he wished for a war of revenge as because he felt that William's Germany was as likely as not to start a war, and that France needed to be prepared.

In all, the period would provide a striking example of the importance of individual talent in the conduct of the affairs of state. Militarily, demographically, and economically, France was hardly any more powerful in the quarter century prior to 1914 than in the years from 1871–1890; Germany was scarcely any less so. Yet the initiative was slipping from German hands and it was in those of France that it largely came to rest. If the diplomatic capital of Europe, for decades, had been Berlin, it now was Paris again, and it was neither guns nor the gross national product that caused the shift.

## THE ROAD TO THE FRANCO-RUSSIAN ALLIANCE, 1890–1894

French intelligence was aided by German error. Within days of Bismarck's fall, the foreign ministry persuaded the Kaiser not to renew the Reinsurance Treaty with Russia, which, unless signed again, was due to expire that summer. The Russians pleaded with Berlin, but the German decision stood. The treaty was allowed to lapse.

One should restrain any easy criticism of the German attitude in 1890. There had been something bordering on the miraculous in Bismarck's performance. Remaining allied to both Austria and Russia was a feat that struck his successors as only slightly less difficult than squaring the circle. The magician was gone now. It might be better to devise a foreign policy that ordinary mortals could manage. What profit had there really been in finding oneself in the middle of every Austro-Russian quarrel?

Besides, the cancellation of the Reinsurance Treaty might well result in attracting a new and rather more desirable ally in the West. The old Chancellor had for some time been trying to draw Great Britain into his alliance system. Now that Germany could no longer be suspected of making common cause with Britain's enemy, Russia, the time seemed propitious to resume the effort. British power was rather more formidable than Russia's in any event, and feelings of kinship a good deal stronger.

There ensued a number of friendly gestures toward Britain. In 1890 came the conclusion of an Anglo-German treaty in which the Germans made some major concessions to the British in East Africa in return for having the tiny island of Heligoland in the North Sea ceded to them. The next year, a grand state and family visit took place as William II traveled to London to pay his respects to his aged grandmother, Queen Victoria.

No matter how sincere those gestures or how valid the motives, the result was a double failure. The Germans had lost one ally without gaining another. The British were not persuaded to join the Triple Alliance; they were chary of cousins bearing gifts. The Russians, on the other hand, faced by the new German coolness toward them, felt humiliated and worried. The old pattern lay broken. So far, after each crisis, they had exploited the facility with which they took diplomatic offense to demand a higher price for returning to their German alliance.

This time, Berlin apparently did not wish them to return, even at lowered rates. The Russians were ready to listen to some other offer.

That other offer came from France. Many people in Europe—some members of the German foreign ministry included—had thought that an alliance between France and Russia was wholly out of the question. A political chasm divided them: France, republican and devoted to the ideals of the Great Revolution of 1789; Russia, tsarist and dedicated to the suppression of even the thought of liberty, equality, and fraternity. But it is not the purpose of an alliance to provide a structure under which governments can grow sentimental over their sympathy for each other or exchange congratulations on the similarities of their views on life, politics, and the universe. Alliances are not friendships. If they were, we would scarcely have any, for national friendships are either illusory or very rare. (National antagonisms are a different matter.) Rather, alliances are in essence concluded so that a country may count on another to fight by its side in times of war, and to support it by a variety of other means in times of peace. A common ideology between allies is a pleasant luxury—it will save the secretary of state from much domestic criticism—but it is not an essential. Mutual advantage is. Common interests, and above all, a common foe, can make allies of nations whose domestic institutions are half a dozen constitutions apart.

These principles certainly applied to the situation between France and Russia. It was true that conflicting systems of government and popular antagonism would delay agreement. But domestic differences would not, in the end, prevent it. With patience, skill, and money, the French government, after 1890, impressed upon the Russians the desirability of cooperating against the common enemy, Germany. The Russians were pleased enough to listen; the French were creating an interesting atmosphere in which to do so. There were French loans to Russia. There was, in 1891, a visit of the French fleet to the naval base of Kronstadt, during which the Tsar, who had made the short trip from St. Petersburg, stood with bared head as the Marseillaise was being played. (Nicholas II would have his day a few years later when the President of France, during another ceremonial visit, had to bare *his* head as the Cossacks marched by.) There were military talks in 1892, cautious but precedent-breaking nonetheless. There was the return visit of the Russian fleet to Toulon in 1893, which set off some enthusiastically pro-Russian demonstrations. There was, finally and most importantly by far, the secret Military Convention signed between Russia and

*The Unlikely Allies? Despite the Franco-Russian Military Convention of 1894 and other indications of cooperation, distrust between the two countries remained, as suggested by this caricature occasioned by the 1908 visit of the French president Clément-Armand Fallières to St. Petersburg, which has Tsar Nicholas II picking the pocket of his French visitor. It appeared in the Paris humor magazine,* L'Assiette au Beurre, *for July 25, 1908.*

France in 1894. Should any member of the Triple Alliance—Germany, Austria, or Italy—mobilize, the two contracting parties undertook to mobilize as well. Should there be war, they would, under certain conditions, fight side by side. These conditions were clearly defined: Should Germany (or Italy supported by Germany) attack France, Russia would give all military assistance to her new ally; should Germany (or Austria supported by Germany) attack Russia, France would come to the rescue.

The secret Franco-Russian Military Convention was a landmark in the affairs of Europe. Three major reasons accounted for that. One was that France no longer was in quarantine; the wall that Bismarck had built had at long last been breached. Gone were the days when France, at every step of the way, was all but automatically outnumbered, outmaneuvered, and outvoted. Another was that the continental armies that would face each other in case of a general war would now be fairly even in size. Not that there was any desire for war. No responsible French statesman, hesitant though he might be to emphasize the point in public, was anxious to provoke a bloody conflict to regain Strasbourg or Metz. Yet—and this was the third reason for its importance—the secret agreement contained an ominous portent. The fates of France and Russia, to quote the eminent American scholar and diplomat George Kennan, had now been linked in such a way "as almost to ensure that there could be no further minor complication in European affairs that would not lead to wider hostilities among the great powers."

## Anglo-French Relations and the Shock of Recognition, 1890–1898

Yet the two blocks into which the Franco-Russian agreement would help to split Europe by 1914 seemed far away. Thus conflict between France and Germany at the time seemed a possibility distinctly more remote than conflict between France and its neighbor across the channel. It was here—between France and Great Britain—that friction seemed most obvious, and most deep-seated. Animosities and resentments older than Joan of Arc set the two countries apart. Modes of life and thought differed profoundly: Glorious Revolution and The Terror, Burke and Robespierre, as seen from the British side; *Roi Soleil* and Cromwell, as seen from the French; Declaration of the Rights of Man and Test Act; and *Coquilles Saint-Jacques* versus fish and chips besides. Nor were differences bridged by joint diplomatic objectives. Instead, conflicting national interests as deep as the Channel divided Britain and France. The century had opened with Great Britain's implacable stand against Napoleon's France. It threatened to end with an almost equally stubborn opposition to Republican France, except that this time, the points at issue concerned colonial rather than continental expansion. England was offended by the idea of further French gains in North Africa; it had largely been with France in mind that she had signed the Mediterranean Agreement. England attempted to block France's ambitions in China, and was willing to fight to prevent any French presence in or near Egypt. Here, of course, was a problem any European nation expanding overseas faced: British interests, holdings, and ambitions were now so vast that wherever on the globe one moved, the British would feel threatened and prepare to do battle. The territorial extent of the British Empire and its lifelines was a credit to British daring, enterprise, and ingenuity. At the same time, it also was an irritant to others, and a source of much international friction.

Trouble between France and England reached its peak in 1898. For some time, both a French-Ethiopian force, under Captain Marchand, and an Anglo-Egyptian army, under General Kitchener, had been heading for the Sudan. (A junior officer in Kitchener's army, by the way, was Winston Churchill, then twenty-four). The background, briefly, was this: The Sudan had been abandoned by the British after a brief period of occupation in the 1880s, when some surprisingly effective local resistance had persuaded them to depart. Since then, the region had been a sort of political no-man's land, fair game, under the Darwinian laws of European imperialism, to the fittest conqueror. The British now were directing Kitchener's army to retake the Sudan at the same time that the French, setting out from Gabon, were moving toward the region in an effort to reach the Nile. In September of 1898, the two forces met at Fashoda, a small town on the Upper Nile.

Wisely, they did not immediately attack. Both forces knew what was at stake. London or Paris would have to give the marching orders. But battle delayed hardly looked like battle denied. Both sides were obdurate. The French would not recall Marchand until the British should make some major colonial concessions elsewhere. The British would listen to no conditions. Kitchener, they said, had retaken the Sudan; let Marchand get out of the way—yield or fight. After six weeks of argument and preparation for war on both sides, the French yielded. Marchand received orders to leave. The British had gambled right. France stood alone, since this was not a conflict in which she could

*The Unlikely Allies? British antipathy toward France—and the readiness even to fight that country if deemed necessary—becomes apparent in this* Punch *cartoon, which appeared during the height of the Fashoda crisis.*

draw on her Russian alliance. France stood ill prepared, racked by the Third Republic's worst *affaire*—the man of the hour was Dreyfus, not Guillaume le Conquérant. And even under different circumstances, it would have made little sense to pour French blood and French resources into the Sudan when Strasbourg was so much closer than Fashoda.

Justify as one might, on sober reflection, the French surrender at first spelled national humiliation. Vituperation of the British in the French press was rarely as unrestrained as in 1898. The Nile, for a moment, did seem to matter as much as the Rhine; the British looked more despicable than the Germans. But in the long run, Fashoda brought about an overdue and salutary reappraisal of French foreign policy. Both Britain and France had statesmen of stature during the affair. On the British side, the Marquess of Salisbury, three times prime minister as well as foreign secretary between 1885 and 1902, had shown his acumen by his correct appraisal of the ultimate decision France would reach, although the bluntness of his demands hardly made that decision any easier. On the French side, Delcassé, who only that summer had taken over the foreign ministry, now asked himself a quite basic question: Could France, which was in no position to fight Britain *and* Germany, afford to hate both countries? His answer was that France could not. It was an unsurprising answer, but it was one that his predecessors had never really thought through; if they had, Marchand would not have been sent looking for the Nile. Henceforth, Delcassé would try his best to reach an accommodation with one of the two, and the power he sought to reconcile was Great Britain.

## Friend or Foe? Anglo-German Relations, 1896–1903

Whether Delcassé would succeed with his new policy was another matter. In France, his decision would be impeded for several years by continuing eruptions of public hostility against the perfidious English. Nor was there much evidence that Britain was eager for any kind of alliance. Had England not fared rather well in splendid isolation, real or imagined? And even if the British should decide to look for companionship, what about the other major contender for their favors—Germany?

The answer to the first question was easier than the answer to the second. Britain, toward the turn of the century, was becoming more receptive toward alliance offers. A great many troubles were plaguing the country; the globe was proving to be too big a place for even Britain to police alone. The Russians were pursuing alarming ambitions in China. The Turks were in trouble again, and there might soon be another San Stefano. France was getting in Britain's way in Asia and Africa and so were other continental powers. Thus the British could do with some help. They decided to have another look at their Teutonic cousin; perhaps his manners had improved a bit.

At first, it hardly seemed they had. The mid-1890s were to see one of the worst instances of Anglo-German hostility on record. It began in Africa. The British, for some time, had had their eye on the gold-rich Boer states in the South—the Orange Free State and the Transvaal. At the end of 1895, Dr. Leander Starr Jameson, an administrator in the British South Africa Company, led a raid from British Betsuanaland into the neighboring Transvaal. He hoped to set off a general revolt against the government of Paul Kruger, President of the Transvaal, which should lead to the replacement of Boer with British control. The raid failed badly. There was no revolt, and the Boers quickly and effectively defeated him and his private army.

Continental opinion—not in Holland alone—was very much with Kruger (David usually has a better press than Goliath); British assurances that Jameson had been acting on his own convinced few. But private feelings were one thing, public policy another, and officially, the powers of Europe pursued a policy of neutrality. All, that is, except Germany. In January, 1896, the Kaiser sent a jubilant telegram to Kruger, congratulating him on his success in repelling the invasion "without appeals for the help of friendly powers." This was a double insult to London, with its joy over Britain's discomfort, and its hint that German aid would have been available to Kruger, and perhaps still was. The telegram was not wholly a matter of personal impulse, as so many of the Kaiser's actions were. For one thing, he wished to impress the British with the need for compensation in the event of further colonial expansion. For another, he may have wished to compel the British into a reappraisal of their German policy—let them see just how isolated they were, so that they might draw the proper conclusions and request admittance to the Triple Alliance.

The British did no such thing, and it is hard to see how they could have. "My dear William," Queen Victoria wrote to Berlin, "I feel I cannot refrain from expressing my deep regret at the telegram you sent President Kruger. It is considered very unfriendly toward this country, which I am sure it is not intended to be, and has, I grieve to say, made a very pointed impression here." Public opinion ran to sentiments similar to the Queen's, though the language was often less Victorian, and demands could be heard for closer ties with Russia and France. The government, in a mood as unsubtle as William's, responded by issuing orders mobilizing a squadron of the British fleet in the Channel. Browbeating people into friendship had its dangers.

But alliance hopes were not dead. People fortunately forget. In the spring of 1898, Joseph Chamberlain, colonial secretary in Salisbury's government, went to see the German ambassador and proposed the conclusion of a defensive alliance to him. The German response was cool. They doubted—correctly so—whether Chamberlain was in fact speaking for the cabinet; nor were they eager to offend the Russians, against whom such an alliance would primarily be directed. Still the idea persisted. The following year, Chamberlain brought it up again, and this time, circumstances seemed more auspicious. Britain was now engaged in a full-scale war against the Boers—the Jameson raid stage was over—and the war was going badly. The Boers were defeating the British, although not forever as it turned out; still it would take the mighty British Empire nearly three years, from 1899 to 1902, to defeat a handful of emigrant Dutch farmers. And long before it was finished, the Boer War would demonstrate two things: the questionable nature of British military power (jibes about the Russian giant as a colossus on feet of clay, for instance, all at once seemed in dubious taste in England) and the extent and intensity of anti-British sentiment in the world. Many were the people who rejoiced with Kruger over every Boer success: the French saw Fashoda avenged; the Russians would take anyone to their hearts, of course, who beat the British; and the Americans remembered another colonial war, and so it went.

The German government, this time, had learned its lesson. Official Germany policy was one of neutrality. No more telegrams to Kruger; instead, toward the end of 1899, the Kaiser paid a cordial visit to England. It was against the background of this visit, then, that Chamberlain made his second plea for friendship. In a speech he gave in November 1899, at Leicester, he called for nothing less than the establishment of a grand alliance between Britain, the United States, and Germany, or as he put it, "a new Triple Alliance between the Teutonic race and the two great branches of the Anglo-Saxon race"; nothing else would do as much "to preserve the peace of the world."

The German reaction resembled the response to his previous initiative. The poor progress of the war in South Africa, argued the people who made policy in Berlin, should before long push the British into making a better, or at any rate a more concrete, alliance offer. Brusquely, they turned down the suggestion.

Undaunted, in 1901 Chamberlain tried once more. Again, he failed. The Germans held out for too much: Britain should join the Triple Alliance; for diplomatic purposes, the Austrians and Italians would have to become Teutons, too. Neither side minded the failure of negotiations—not then. The British thought that cooperation with Japan might remove some of the immediate pressures in the Far East, and in 1902 concluded an Anglo-Japanese alliance directed against further Russian expansion in the area. The British also made the most of their new cooperation with the United States, which depended on no formal alliance but was, nevertheless, a fact of diplomatic life ever since the Spanish-American War had catapulted the United States into world politics. The Germans still thought that they had all the time in the world, reasoning that Britain had no choice; surely, she would never reach an agreement with her arch-rivals Russia and France. Fashoda and the Crimean War made comforting memories. No, Britain would come back and make a more handsome offer.

This brings to mind Fontenelle: "Leibniz never married. He had considered it at the age of fifty, but the person he had in mind asked for time to reflect. This gave Leibniz time to reflect, too, and so he never married."

The British proposal, quite possibly, was no more genuine than Leibniz's. What really were they offering the Germans? They never said. But why did the Germans not test the British offer by saying yes? The British move may have been a matter of tactics rather than strategy, but tactics have a way of developing a momentum of their own. At any rate, the British, having had time to reflect, would not woo the Germans again. The balance sheet of the affair makes for melancholy reading.

Only one thing remained on (or rather near) the positive side: this was an Anglo-German Agreement, concluded in 1898, to divide Portugal's African colonies in case Portugal should, as was then expected, find itself in serious financial difficulties and have to pawn parts of her empire. Actually, it was a perfectly useless treaty. The following year, the British renewed their ancient alliance with Portugal. Although the evidence for this is circumstantial rather than direct, they also appear to have provided Lisbon with substantial amounts of pounds sterling; quite suddenly, the country's finances improved, and Angola and Mozambique would remain Portuguese for quite some time after most other powers had departed from Africa.

On the negative side, Anglo-German friction would continue nearly unchecked. To begin with, there was the naval race. Starting in 1898, the Germans began building a major navy, arguing that they now had overseas possessions to supply and defend. Continental power no longer sufficed. "Our future," said the Kaiser, "lies on the water." The British, in alarm, reacted by increasing *their* naval building program, whereupon the Germans, after a suitable interval, increased theirs again. And both governments found that the most effective method of persuading their respective parliaments to pass the necessary appropriations was to dramatize the threat from the other side. It was not an atmosphere conducive to friendship.

Then there was the matter of colonial rivalry. Even Bismarck had not been able to hold out forever against the dynamics of imperialism. In the mid-1880s, the Germans began, modestly at first, to acquire African territory. Under William II, all restraints were off: Germany must become a great colonial power! Bismarck's insistence that Germany was a satiated nation was no longer heeded—Germany must show the world that her energies and achievement were second to none. A great power must grow or die; few half-truths have caused as much misery in the world as this.

There was, finally, the matter of economic rivalry with England. The Germans were ambitious to outsell the British in the world's markets, and to redraw some economic spheres of interest as well. In 1899, a German company received a concession to build a railroad from Constantinople to the Persian Gulf, the "Berlin-Baghdad Railway." The project was more grandiose than the available financing, and for monetary as well as diplomatic reasons, the Germans urged the French and British to participate. But by 1903, that effort had failed, since neither the French nor the British would take part. The Russians had fumed about the project from the beginning; they felt that by some ancient right, this was their part of the world. In the case of the French, anti-German sentiment was still something of a reflex action. But that the British, too, should be so dubious about the Germans' motives was a poor omen.

It would be easy to add to these instances of Anglo-German differences. But this is not an encyclopedia of diplomatic crises. Besides, such an unrelieved recitation of troubles can falsify what took place, obscuring the amount of good will that still did exist between the two nations. To give but one instance, the Rhodes scholarships endowed in this same period restricted the recipients to citizens of Great Britain, the United States, and Germany—precisely the combination Joseph Chamberlain had had in mind. And it makes events appear far more inevitable than in fact they were. Many of the points at issue were fairly casual matters at the time. Too many other cares and enmities loomed larger. The British might worry about the Berlin-Baghdad project, but compared with, say a hypothetical Moscow-Baghdad Railway it was a minor care. It is only in retrospect that the various instances of Anglo-German friction look so momentous. Hard as it is to avoid in the writing of history, *post hoc ergo propter hoc* remains one of the most basic of fallacies. Still, one consequence of it all was that unlike Leibniz, the British now were willing to try marriage—but with a different partner.

## Entente Cordiale, 1904

In 1900, Lord Lansdowne had succeeded Salisbury as British foreign secretary. Salisbury had held out, with Washingtonian conviction, against any entangling alliances. Lansdowne did not. It was not that he was any fonder of alliances as such. It was rather that he had taken a new and sharp look at certain facts. What he saw, in the first place, was the far from splendid sort of isolation in which England found herself. What he also saw was that the major problems of the period were not continental but colonial. It was overseas, in Africa and Asia, that the principal pressures and conflicts were accumulating. Had it been otherwise, England would have had no cause to abandon nonalignment, since in Europe, she held precisely what she wanted: the perfect balance of power between the two rival alliances, the Franco-Russian, and the Austro-German-Italian. But overseas, in the pursuit of imperialism, England faced too many enemies; she also faced the danger that these enemies—Russia, France, and Germany—might one day unite against her. Britain, Lansdowne felt, would have to come to a settlement with at least one ally.

But which one? Russia seemed irreconcilable, her appetite limitless: Manchuria, Korea, Persia, the Dardanelles, and who could tell what else the Tsar might want. Germany made a better, and certainly a more likeable prospect, but the Germans had just demonstrated their reluctance to come to an agreement. This left France. Here, then, was the opportunity Delcassé had been waiting for, and he made splendid use of it, showing now exactly the sort of moderation, patience, tact, and comprehension of essentials that the Germans had so lacked during the three Chamberlain episodes. And mood and circumstance were right. It was the period of Edward VII, who, since succeeding Queen Victoria in 1901, had done a singular job of dispelling a major portion of anti-British feeling in France by his obvious enjoyment of things pleasantly French. (Also, Lansdowne, on his mother's side, could claim descent from Talleyrand, but that fact may deserve no more than parenthetical mention. Sir Eyre Crowe, one of the senior permanent officials in the foreign office and its most persistently, almost hysterically, anti-German member, was half German.)

In the spring of 1904, negotiations, begun the year before, were complete. Formally, the agreement reached consisted of a number of separate and concrete accords; informally, it would become known in its entirety as the *Entente Cordiale*. And "Cordial Understanding" was a good term, for the two countries were agreeing to settle all outstanding colonial issues between them, minor ones and major ones alike, from disputes over customs duties in Madagascar to fishing rights off the coast of Newfoundland. (The latter, by the way, proved to be more resistant to resolution than any other single problem discussed.) But the core of the argument dealt with Egypt and Morocco. France agreed to give England a free hand in Egypt; the British, in return, agreed that Morocco should be in the French sphere. There were to be no more tests of strength and temper over ill-defined regions of interest—there were to be no more Fashodas.

Was the alliance directed against Germany? Was it, in fact, an alliance? The negotiators involved could say no to both questions, though not with an altogether easy conscience. It was true that Germany was not mentioned and that the *Entente* ostensibly dealt with colonial, not with European affairs. It was also true that there were no provisions for assistance in case of war, the usual characteristic of a major alliance. Yet the risk was there that the Moroccan provision might involve England on the side of France, since the argument over Morocco could—and would—as easily be one between France and Germany as between

PUNCH, OR THE LONDON CHARIVARI. August 2, 1911.

SOLID.

GERMANY. "DONNERWETTER! IT'S ROCK. I THOUGHT IT WAS GOING TO BE PAPER."

*Despite Anglo-German sympathies, and Anglo-French or Franco-Russian antipathies, the new power grouping held. Bismarck's system was dead.*

France and England. Above all, Britain *had* emerged from isolation, and no matter how cautious and limited the documents that attested to that fact might be, she had come out on the side of France, not of Germany. Time now would tell whether the *Entente*—as Delcassé but perhaps not Lansdowne wished—would indeed assume the character of an alliance.

## Testing the Entente: Russo-Japanese War, Moroccan Crisis, Björkö, and the Matter of Staff Talks, 1904–1906

For the moment, it appeared as though the *Entente* might not survive at all, let alone develop into a firm alliance. Within six months of its conclusion, England went to the edge of war with France's eastern ally, Russia. The setting was provided by the Russo-Japanese War of 1904–1905.

The rivalry between Russia and Japan was old and bitter; both were seeking to profit from the decline of China by expanding into Manchuria and Korea. Early in 1904, the Japanese chose major war over minor skirmishes. Without warning and without a declaration of war, they attacked the Russian fleet at Port Arthur. The Russians, outfought and outthought, decided to bring up their reserves: they would send their Baltic fleet half way around the world to defeat the Japanese. On the lengthy journey to the Far East, that fleet won one victory. Shortly after starting out from port in October 1904, it encountered what looked to its officers like a group of Japanese torpedo boats near the Dogger Bank in the North Sea. The Russians opened fire, missing many, but managing to sink one. The pity of it was that Russian marksmanship was even this good, for the Japanese men-of-war in truth were British fishing boats.

The British were incensed. Suddenly the two countries were as close to war as they had been at the time of San Stefano. It was a prospect that, elating though it might be to many a Britisher and Russian, horrified Delcassé. Using all his powers of persuasion, he kept France's two allies from each other's throats. Grudgingly, the Russians admitted their error, promised compensation, and even put the responsible officers ashore. The rest of the fleet, minus the myopic officers, sailed on, to be annihilated by the Japanese at the battle of Tsushima.

Soon after that battle, in the fall of 1905, peace was negotiated on neutral territory in Portsmouth, New Hampshire. (A considerable part in arranging the negotiations had been played by President Theodore Roosevelt; his successful mediation would bring him the Nobel Peace Prize in 1906.) Considering Russia's performance in the war, the peace terms were lenient; even so, Portsmouth put a temporary stop to Russia's ambitions in the Far East.

The consequences of the Russo-Japanese War for European diplomacy were twofold. In the first place, the Dogger Bank incident marked a crucial point in Anglo-Russian relations. Had the British retaliated, as easily they might, there would be many a learned volume now explaining how the great Anglo-Russian War of 1904 was a logical and inevitable development of decades of friction between London and St. Petersburg. And indeed, if some sort of apparatus for the precise measurement of hostility between nations were to exist, Anglo-Russian relations, in the half century or so before 1904, might receive its highest score. But, narrowly or not, war had been avoided, and the two countries found that they could talk as well as shout, particularly when they could draw on the offices of as good an intermediary as France. The second consequence, connected, of course, with the first, was that Russia was temporarily in a chastened mood. The direct domestic result of the bungled war was the Revolution of 1905. Tsardom would have a difficult enough time merely to survive. Thoughts of expansion, at least in the Far East, would have to wait. The attention of Russia's policy makers, once the immediate crisis was over, turned to Europe again, an attitude that did not at all displease the French.

But how strong actually was France's understanding with England? The Germans decided to find out, now that the happy prospect of a battle between France's eastern and northern allies had vanished. In March 1905, the Kaiser, on a Mediterranean cruise, landed at Tangiers in Spanish Morocco, and publicly declared Germany's support for Moroccan independence and integrity. Diplomacy's public language is often circuitous, and William did not, with these words, mean to express any true support for Morocco's native regime, which was hopelessly ineffectual. What he did mean to suggest was that Germany had as much of a claim to Morocco as any other greedy European power. To make that point even clearer, the Germans followed up the Imperial speech with a demand for an international conference on Morocco. The French, they calculated, were sure to be outvoted in such a conference, in which Germany would demand an Open Door policy for Morocco. Spain and Italy, they reasoned, were sufficiently jealous of France to vote with Germany; Russia was exhausted and dependent on Germany's good will; Austria-Hungary was Germany's firm ally; and the British were anxious to safeguard their profitable trade with Morocco and thus to favor a policy of the Open Door. Even if Germany's direct gains from such a conference should be slight since there were too many contenders for Moroccan privileges, the diplomatic isolation of France would once again appear as plainly as in Bismarck's days.

The French, considerably worried, went out of their way to suggest some compromise. In May, the French government offered the Germans a port on the coast of Morocco. In June, the cabinet dropped Delcassé, who alone was holding out for a hard line against Germany. The Germans remained adamant; they wanted their conference. Their wish no longer could be resisted, and early in 1906, in the Spanish town of Algeciras, the conference convened. But the Germans had pushed their luck too far, and hardly anything worked out the way they had expected. The British were very much disturbed by what they rightly took to be a case of German bullying. The Russians were much more intrigued by what the British rather than the German delegates had to say. The Americans showed a decided preference for accents that reminded them of Lafayette rather than of the Hessians. The Spaniards and Italians rediscovered their Latin bonds as they listened to French references to compensation, a topic the Germans had not had the wit to pay much attention to. When the votes were counted, the Germans were in a minority in every instance; none but Austria-Hungary would side with them. Except for its relatively small Spanish section, Morocco, in effect, was to be open to French economic and political penetration. Even more important than the technicalities of the settlement was the cooperation between the French and British delegations at Algeciras. The Germans had meant the conference to be a test of the *Entente*. It was, except that the results were all wrong. Germany, not France stood in isolation. The *Entente* had emerged stronger than before.

Overplaying a hand remained a German specialty. In the early stages of the Moroccan crisis, in July 1905, Tsar and Kaiser, pleasure-cruising in the Baltic, met near the Finnish island of Björkö. William, at the time, was still savoring Delcassé's fall; Nicholas was still brooding over the humiliation of the Dogger Bank episode. The Kaiser, visiting the Tsar's yacht, brought along the draft of a treaty, which, he told Nicholas, "I happen to have quite by chance in my pocket." The treaty provided for a defensive alliance between Germany and Russia against attack from any other power in Europe. In other words, it was a treaty that would have neutralized Russia as France's ally. How would his cousin Nicholas like to sign it, right then and there? "It would be a very nice souvenir of our get-together." Nicholas did want to sign it. "Bright tears of joy" wrote William, "stood in my eyes." Just like that, Germany and Russia were allies again.

But not quite just like that. When the two sovereigns returned to dry land, they were astonished by the political reception to their action. The German Chancellor, Prince Bernhard von Bülow, at first would have no part of the agreement. He had no objection to a treaty with Russia, but this was not the form he had in mind. Bülow was no model of either statesmanship or courage, but this particular instance of Imperial casualness was too much even for him. Only a desperate plea from William—"think of my poor wife and children"—kept him from resigning the chancellorship. Thus Bülow decided he would make the best of a bad opening move and let the treaty stand. It was a pointless decision, because in Russia, the Tsar had meanwhile had to listen to his foreign ministry's obvious argument that the Björkö agreement plainly violated the spirit of the French alliance. With some embarrassment, he wrote the Kaiser that the treaty could not go into effect quite yet. Also, might it not be possible to add a few words stating that the agreement was to be suspended in the event of war between France and Germany?

Of course, no such addition was possible. "What is signed is signed and God is witness of it!" replied William. His protest went unheard, and the treaty never did go into effect. All that remained were injured feelings. "Two great princes who wish to live in amity," the French diplomat and historian Philippe de Commynes had written nearly half a millenium earlier "should never meet, but send each other good and wise envoys."

The French now showed they had retained a great deal of Commynes' good sense. At the end of 1905, while the Moroccan situation was still critical and peace by no means certain, they asked the British for a promise of military support in case of war with Germany. Sir Edward Grey, who at just that moment had succeeded Lansdowne as foreign secretary (a post he would hold until 1916, longer than any other foreign secretary ever continuously held office) did not feel that this was a promise he had either the inclination or the authority to make. What he did do, however, was to authorize informal military conversations

Die Monarchenzusammenkunft in der Ostsee

„Zeigen Sie uns mal, wie man rückwärts steuert — das interessiert uns am meisten."

*"The Monarchs Meet on the Baltic" is the heading of this cartoon in the German satirical weekly* Simplicissimus, *with Nicholas II on the left and William II on the right. The caption reads, "Show us how to go into reverse, will you—that's what interests us the most." (The artist is Thomas Theodor Heine, one of the magazine's principal cartoonists and founding members.)*

between the British and the French, conversations that began in 1906 and continued until Sarajevo replaced them with other arrangements.

No binding agreements were ever signed in the course of these talks, no pledges of aid exchanged. In fact, the British cabinet would not even learn of their existence until 1912, the British public not until 1914. However, the military experts involved did agree on a number of very specific details of military cooperation, *if* such cooperation should ever be called for, from the size and deployment of a British expeditionary force to an understanding concerning the most efficient use of both fleets—the British were to guard the North Sea, the French the Mediterranean. Theoretically, then, both nations preserved perfect freedom of action. Actually, England's moral obligation to France increased as the talks went on. The British could not very well ask the French to deploy their navy in the Mediterranean without assuming *some* implied responsibility for intercepting a German fleet heading for Dunkerque or Cherbourg. Nor was it really feasible to carry on, over a period of eight years, talks involving some of the most precise details of possible military cooperation without giving the French—the phrase is Grey's—"an expectation of support."

In part through German clumsiness, in part through French ingenuity, sometimes by gradual stages, sometimes by sharp crises, the *Entente* had turned into something startlingly like an alliance.

## TRIPLE ENTENTE: THE ANGLO-RUSSIAN UNDERSTANDING OF 1907

The trend from *Entente* to alliance suggested to the British the wisdom of reassessing their relations with Russia. Avoiding friction with one's ally's ally is a fairly basic tenet of statecraft. Besides, there was always the fear that the next time Tsar and Kaiser met, they might come up with another, better thought out, Björkö agreement. But that was not the side Britain wanted Russia to be on. Considering Germany's apparent designs on the Near East, the British could use an ally who was in a position to oppose the prospective builders of a railroad from Berlin to Baghdad. Moreover, now that Russia had been expelled from China, and thus some definite points of Anglo-Russian friction had been reduced, the possibilities of agreement between London and St. Petersburg looked better than they had for a long time.

Encouraged by the French, the British, after Algeciras, urged the Russians to consider the advantages of a *rapprochement*. The Russians acted reluctant. The Tsar was preoccupied with the problem of the regime's domestic survival, and the revolution of 1905 was still a bad and vivid memory. He also was concerned lest he give offense to the Germans. Britain might have the world's greatest navy and shiniest gold reserve, but Britain was far away. Germany was a neighbor. But then the British held out a lure too sweet to resist. They might, just might, change their views on the Dardanelles and favor the opening of the Straits to Russian shipping in war as well as in peace. (The Russo-Japanese War had just demonstrated again how galling the provision was which closed the Straits in the event of war. Russia's Black Sea fleet had been imprisoned by the Straits, and hence wholly useless.) In truth, the British promise was vague, but the Russians chose not to read it that way. In 1907, they signed a formal agreement with Great Britain. The agreement was similar to the Anglo-French *Entente* of 1904 in that, on the face of it, the two powers were merely settling certain outstanding specific colonial issues between them rather than entering into a military alliance. It was similar, too, in that the whole of the agreement was really greater than the sum of its parts.

Specifically, what the Anglo-Russian accord of 1907 settled was the fate of Tibet, Afghanistan, and Persia (it still bore that name then rather than that of today's Iran). Both sides agreed to stay out of Tibet, while Britain was to have a free hand in Afghanistan. Persia was divided into spheres of influence: the North went to Russia, the South to Britain, and a neutral zone in the middle provided a buffer between the two. It was imperialism at its bluntest as well as at its most hypocritical: "a mutual self-denying ordinance recognizing Persian independence," Grey piously called it.

No matter; what counted was that Britain and Russia had agreed to settle their principal points of imperialistic conflict by negotiation, as Britain and France had done three years earlier. And amicable and pacific as the agreement was, there was implicit in it a considerable element of political danger. Hitherto, Russia's ambitions in the Balkans and in the Near East had been restrained by the powers, notably by England. Now, Russia had reason to assume that she might some day count on Western aid and comfort in the pursuit of her aims. Beyond that, what some writers have called a diplomatic revolution had taken place. The Triple Alliance of Germany, Austria-Hungary, and Italy (or rather the Dual Alliance, for Italy's reliability had become a matter of justified doubt) after 1907 found itself faced by the Triple *Entente* of France, Great Britain, and Russia. It was the triumph of France. Utterly alone in 1871, a generation later she was *à trois* in a world ruled by five powers. To quote Bismarck again, it was the kind of union which "would provide the basis for a coalition against us more dangerous than any other which Germany could possibly face."

## Bosnia and the Annexation Crisis of 1908–1909

The country that initially courted conflict with the new coalition was not Germany, however, but Austria-Hungary. In 1908, the Austrians thought that they might follow the first step they had taken in 1878 with a second. (When the Austrians talked about deliberate speed, the stress was on the adjective.) They would annex the provinces of Bosnia and Herzegovina which they had been given to "occupy and administer" at the Congress of Berlin. There were several good reasons for the annexation—for instance, the need to clarify the legal status of Bosnia-Herzegovina, and the desire to demonstrate to the Turks that returning the provinces was out of the question. But probably the strongest single reason was that held by Count Lexa von Aehrenthal who two years earlier had become foreign minister, and who felt very strongly that Austria, to survive, needed to show that she was still capable of action.

Aehrenthal, while ready to pursue a policy that contained its share of risks, was not a reckless man; he would first probe the Russian reaction. In September 1908, he met his Russian counterpart, Alexander Izvolski, at Buchlau in Moravia. Izvolski was by no means delighted at Aehrenthal's plans, but he was willing to strike a bargain: Russia would not oppose annexation if Austria, in return, would support the opening of the Straits to Russian warships. Aehrenthal agreed, although by design or not, he volunteered very little information about the precise date of Austria's intended action on either Bosnia or the Straits.

A month later, in October 1908, the Austrians proclaimed to the world that they were annexing Bosnia and Herzegovina. Izvolski was furious. He felt that Aehrenthal had broken his word: it had been his understanding at Buchlau that annexation was in the fairly distant future and would follow rather than antecede a new Straits settlement. Angrier even than the Russian foreign minister were the Serbians. They had had their own designs in this area; the region's citizens were fellow

Slavs. Crowds in Belgrade called for war. Their government seemed willing to oblige them.

Britain and France rallied to support their Russian ally. Austria's action was a violation of the Treaty of Berlin and should be opposed—but not to the point of war. This, on sober thought, was the way the Russians reacted too, since they knew they were in no position to fight a European war this soon after the debacle of 1905. They also knew that Austria could count on German aid—reluctant aid, to be sure, for Berlin had had as little advance warning of Aehrenthal's action as anyone else, but aid nonetheless. Thus, galling though the task might be, St. Petersburg intervened to restrain Belgrade.

Ultimately, when the crisis was over in 1909, war had been averted, but hardly anyone had gained. A Russian attempt to use the crisis to obtain concessions on the Straits foundered on British opposition. The ties of the Triple *Entente* were not that firm; neither its letter nor its spirit provided for British support of all of Russia's national aspirations. St. Petersburg's firmer ally Serbia, would not forget nor forgive 1908, disliking the Russians for their lack of support, but hating the Austrians even more. Austria, it was true, managed to keep Bosnia-Herzegovina, but the price, a cash compensation to Turkey aside, was Russian enmity. Those Austrians, and there were many, who had argued against changing the status of 1878, legally awkward though it might be, had been right. Less ambition on Aehrenthal's part—Talleyrand's *pas de zèle* "go easy on the zeal"—would have served Austria better.

## The Panther's Leap: The Second Moroccan Crisis of 1911

A period of relative calm and of a search for accommodation followed the annexation affair. Not all was a blind rush toward catastrophe, of crisis pursuing crisis in pre-1914 Europe, nor was what had taken place in 1908 and 1909 that corrosive to international good will. Tranquillity is harder to record than trouble, of course, and general and genuine accommodation would remain elusive. There would still take place, in the five years between Aehrenthal's coup and Princip's shots, one major crisis and several minor wars. Any serious diplomatic narrative must mention them, but it should also note, even if in this cursory fashion only, the existence of normalcy, peace, and a sense of proportion on the part of many of Europe's statesmen.

The major crisis involved Morocco. The Algeciras conference had given France considerable rights in Morocco, although these rights were hedged with provisions that recognized the continuing economic interests of others in the region. In 1911, when a native revolt led to open French military intervention, the Germans took the opportunity to remind Paris that Morocco was still nominally independent, and that they could set some limits to French expansion. Early in July, 1911, the German gunboat *Panther* appeared at the Moroccan port of Agadir, ostensibly to protect German citizens there, but actually to promote much larger German ambitions, although these did not involve Morocco itself. The Germans had no designs there. What they did want was compensation elsewhere. They would, they confided to the French, be willing to recognize a French protectorate over Morocco if in return they were to receive the entire French Congo.

The demand was immoderate, the method childish. Gunboat diplomacy is a game for experts, and the Germans were amateurs. The "*Panther's* leap," as the press came to see it, gave German diplomacy an image of crudeness and brutality. It also caused a great deal of alarm in Britain, as the British dreaded the thought of seeing the Germans this close to Gibraltar. The Kaiser, they thought (wrongly, by the way) meant to turn Agadir into a German naval base. Thus on July 21, 1911, David Lloyd George, in the course of a speech he gave at London's Mansion House, repaid the Germans in coin as undiplomatic as that they had used themselves.

Lloyd George, Chancellor of the Exchequer since 1908, was in a bad mood as far as Germany was concerned. He had long been an advocate of a whole series of social reforms, from health insurance to old age pensions, reforms that, no matter how socially desirable, cost a great deal of money. Yet a disproportionate amount of the British budget had to be allocated to naval armaments. The Germans were engaged in a major program for building dreadnoughts—the newest of the "all-big-guns" battleships—in which Britain could not afford to be overtaken. Anger at both the implications of the naval race and at German heavy-handedness over Agadir found an outlet then in Lloyd George's Mansion House speech, when he said:

> I believe it is essential in the highest interests, not only of this country, but of the world, that Great Britain should at all hazards maintain her place and prestige among the Great Powers of the world. . . . If a situation were to be forced upon us in which peace could only be preserved by the surrender of the great and beneficent position Britain has won by centuries

> of heroism and advancement, by allowing Britain to be treated where her interests were vitally affected as though she were of no account in the Cabinet of nations, then I say emphatically that peace at that price would be a humiliation intolerable for a great country like ours to endure.

This was not the public language of diplomacy, especially in a case where no direct British interest was at stake, and the Germans had some right to be offended. But British support for France was clear, even if in subsequent talks the tone was more genteel. This part of the *Entente* was holding. The Germans chose to come to terms. In November 1911, after a great deal of difficult bargaining, agreement was reached: the French ceded some worthless strips of territory in the Congo to the Germans; in return, the Germans gave France a free hand in Morocco and withdrew their naval units.

Again, as in the Bosnian crisis, it was a solution short of war, but one short of giving general satisfaction as well. Many Frenchmen were displeased because a part of their colonial empire had been surrendered to the Germans. Many Britishers were supplied with new grounds for disliking their transaquatic cousins. Many Germans thought the Congolese gains hardly worth the risks and trouble taken, and right they were, for another parallel needs to be drawn between the Bosnian and Second Moroccan Crises: both involved prestige politics at its very worst. Neither Austria's position in Bosnia, nor Germany's in North Africa, by any reasonable calculation were worth a war, and no power, in fact, seriously contemplated war. (Serbia, in 1908, was the exception to the rule.) Yet war was talked about much too freely. Thus, through a combination of maladroit diplomacy, ill-advised publicity, and entangling alliances, minor matters were turned into major issues of national pride and international concern. Still, Europe had survived both affairs, and the next test of nerves would come from a quite different source and, incidentally, would prove that part at least of the new alliance system was still quite capable of being disentangled.

## The Troubles of Turkey

### *The Tripolitanian War, 1911–1912*

The different source first was Italy, and after that four Balkan nations. Rome's appetite was growing with the French eating. The Italians, after watching the success of France in Morocco, decided in 1911 to act on an old plan of theirs. They would annex Tripoli. Tripoli, aside from the fact that the Turks who held it were weak, had little to recommend it. Its resources were negligible, its inhabitants quarrelsome, its soil largely desert. It was, in sum, such an undesirable area that neither had the French cared to expand into it from Tunisia, nor the British from Egypt. If it had not been a matter of keeping up with the Schmidts, the Dubois, and the Joneses, it is hard to see what could have attracted the Italians to Tripoli.

But the game of keeping up was precisely the one the Italians wished to play, and they had prepared the diplomatic ground very carefully. They had long ago informed most major powers of their intentions and purchased their approval by various means. In the case of France, the Italians had, as early as 1902, agreed to remain neutral in any war in which France was the aggrieved party, to which they had added an accord that as Tripoli was to be within Italy's sphere of influence, so Morocco was to be within that of France. In the case of Russia, the equally secret Racconigi Agreement of 1909 (signed at that small Piedmontese town near Turin) provided for a comparable trade: Italy would back the Russian position on the matter of the Straits; Russia would support Italy's designs on Tripoli. *Quid pro quo*, although the delicate question remained of how these agreements might be reconciled with Italy's professed adherence to the Triple Alliance. Here was Italy, bound under the terms of the regularly renewed Triple Alliance of 1882 to come to the aid of Germany and Austria, concluding secret agreements with the potential enemies of her allies and, in the case of France, very clearly promising neutrality.

There was, the Italians knew, no way of reconciling the two sets of obligations. Italy was changing sides, and moving into the French camp.

A good argument could be made for the shift. The Habsburgs hardly were Italy's natural ally, and the French, who were present in North Africa, were in a position to offer more than the Germans, who were not. Alliances are not eternal; public assurances of national friendship need be no more sincere than the word "sincerely" used in closing a letter. Even so, there remains something shabby about Italy's maneuvers. In mood and essence, they were too naïvely Machiavellian. Everything in *The Prince* clearly depends on the manner in which it is executed, and the Italian approach lacked style and elegance. Whoever offered Italy more sand could have her in partnership. The sad fact was that prewar Italy missed Cavour's talents as badly as Germany missed Bismarck's.

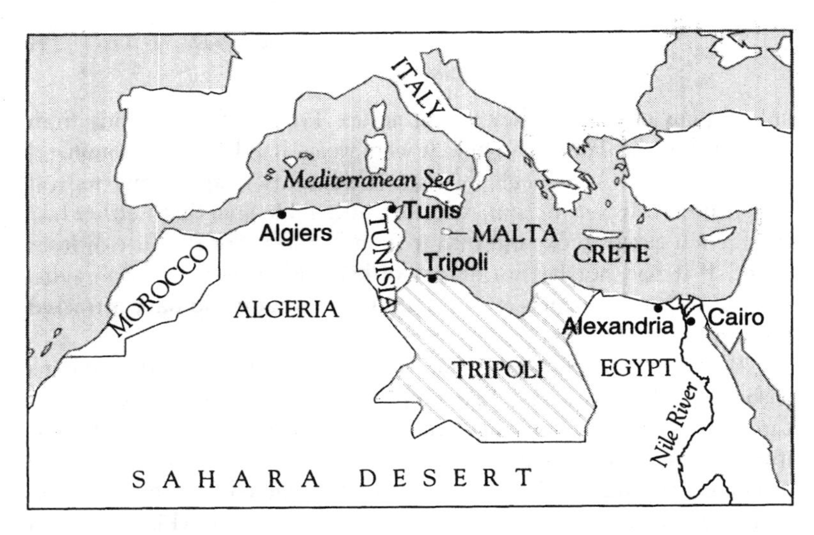


*Tripoli, ca. 1912.*

The Balkan Wars, 1912–1913. *In 1912, Bulgaria and Serbia, in association with Greece and Montenegro, fought the Ottoman Empire to detach its last remaining Balkan possessions, those of Macedonia and Albania. A year later the Second Balkan War saw Bulgaria facing its recent allies, who were now joined by formerly neutral Rumania.*

For the moment, however, there was still room for illusions. The Italians did well in the Tripolitanian war; they were gaining what they had set out to acquire. (It should be noted, however, that not all Italians were enthusiastic about the war with Turkey. Among those most violent in opposing it was a Socialist newspaper editor who, in the course of his public campaign against it, called the national flag "a rag, to be planted on a dunghill" and who soon thereafter found himself in prison for this particular contribution to the war effort; his name was Benito Mussolini.) By 1912, the Turks acknowledged that they had lost the war, and formally surrendered Tripoli to the Italians, who renamed it Libya, a change that would neither make the natives more friendly nor the soil more fertile.

It was not until later that oil would be discovered in Tripoli-Libya, and the country would grow very rich, rich enough in fact to become a major financial supporter of international terrorism. But that was in the future, and even if the presence of oil had been known at the time, it would have made little difference, since oil was not yet high on anyone's list of vital resources. Thus, as seen from the contemporary perspective, Tripoli was one of the most clearly unprofitable of the ventures undertaken by any of the European powers.

## *The Balkan Wars, 1912–1913*

Even so, it was also one of the most contagious, for the ease with which the Italians had won their victories led a number of Balkan states to decide on taking action of their own against Turkey. In 1912, Bulgaria, Serbia, Montenegro, and Greece forgot their mutual quarrels long enough to unite for war against Ottoman rule in the Balkans. The Russians, to whom their enemy's enemy was a friend, were happy to encourage them. Relatively little urging was required. The stakes were high. Macedonia and Albania, all that remained of the Turkish imperial presence in Europe, might fall to the attacker. The moment seemed propitious. The Italians had opened the gates, and it would be wise to follow them in without too much delay. They might swing shut again, since not long before a disquieting event had taken place in Constantinople. A group of Young Turk reformers, eager to bring their nation into the modern age, had seized many positions of power in 1908, and it seemed essential to move before their reforms had a chance to go into effect. (It was a thought that had played its part in the Austrian annexation of Bosnia as well.) Turkey, as its poor showing against Italy

*A British comment on Italy's adventure in Tripoli. It perhaps unintentionally reveals a curious blind spot in British thinking which would plague the country's relations with much of the rest of the world. It was a tendency to be touchy about the ambitions of others, without apparently realizing that it was Britain that had taken the lead in the race for overseas expansion, and in the process had accumulated by far the largest colonial empire of any European power.*

demonstrated, was still satisfactorily weak, and the Tripolitanian war, while coming to a close, was still engaging Constantinople's attention on another continent. This, then, was the time to strike.

Militarily, the four Balkan allies did exceedingly well in their war against Turkey, defeating their enemy wherever they engaged him. Diplomatically, they did less well. The powers were agreeable to letting them divide Macedonia, but Austria successfully objected to allowing Serbia and Montenegro to acquire Albania. To give Belgrade this much new territory and access to the sea as well, thought the Austrians, would make for too frightening an increase in Serbian power. Still, it was victory. When the peace treaty was signed in May 1913, Turkey had ceased to be a European power.

Within a few weeks of the peace, normalcy returned, as the Balkan nations fell to bickering with one another again. None thought the division of the spoils a just one. At the end of June 1913, the Bulgarians resorted to direct action and attacked the Greeks and Serbians. It was an unwise move. Within a very short time, they were outnumbered five to one. Montenegro, allied to Serbia, joined the anti-Bulgarian side. So did Rumania, hitherto neutral, but now scenting the chance to recover the Southern Dobrudja, which had been accorded to Bulgaria at the Congress of Berlin. So, for similarly non-philanthropic reasons did Turkey, yesterday's enemy of Serbia and Greece. In August 1913, the Second Balkan War was over, and the Bulgarians were not the winners. Under the terms of the Treaty of Bucharest, which formally ended it, much of the Macedonian territory claimed by Bulgaria went to Greece and Serbia, while the Dobrudja went to Rumania. Albania escaped undivided. Instead, it was established as a nominally independent state, which would become celebrated, for years to come, for the quality of its political instability.

Why retell this story of the two Balkan Wars, confusing even in this barest outline? Not because they were of vast direct significance in the affairs of the major powers. Balkan quarrels were not that crucial. It is told again, rather, because it contains some of the germs of 1914: the expansionist ambitions of the new nations of Southeastern Europe; the unashamed use of blunt force to settle territorial conflicts; the encouragement, on Russia's part, of Balkan violence; the opposition, on Austria's part, to Serbia's unchecked growth; the dissatisfaction of Bulgaria with the Macedonian settlement. None of these, to reiterate the point, were matters of cosmic import. Yet all of them would be a part of the background to the crime of Gavrilo Princip. Thus,

**Der Brand am Balkan** (Zeichnung von Th. Th. Heine)

Der vereinigten europälschen Feuerwehr gelang es leider nicht, den Brand zu löschen

*War in the Balkans. The inability of the rest of Europe to put a halt to the outbreak of hostilities in the Balkans is illustrated by this cartoon in* Simplicissimus *by Th. Th. Heine. It shows the four powers—symbolized by the lion for Britain, the rooster for France, the bear for Russia, and the helmeted eagle for Germany—under the heading of "Fire in the Balkans," with the caption reading, "Regrettably, the United European Fire Department was unable to extinguish the blaze." (What the cartoon also shows is that for all the points of friction between them, the powers of Europe still had some interests in common, even if Heine was perhaps too charitable about Russia's role.)*

1914–1918 would merely be the largest, but by no means the only one, in a series of wars of the Turkish succession.

## Road to Disaster? 1871–1914 in Perspective

The Balkan quarrels were a *part* of the background, and, in this case, the qualifying phrase is more than just another instance of the historian's tendency toward automatic equivocation. For no solution of Balkan problems can be imagined that would have pleased everyone concerned and that would not have contained some seeds of future conflict. Each of the rival aspirations—Bulgaria's, Serbia's, Montenegro's, Rumania's, Greece's, even Turkey's—had its historic or ethnic justification. Claims and rights were so contradictory and so entangled that the wisest and most disinterested of judges imaginable could never establish one side's justice except at the cost of another's injustice. No matter what the solution, resentment would flourish. Tragedy, in the Hegelian definition, is not the conflict of right with wrong, but of right with right. This, indeed, was one of the basic aspects of Southeastern Europe's tragedy. Hence, while the aftermath of the Balkan Wars—with Serbia anxious for further aggrandizement, Bulgaria impatient to overturn the verdict of Bucharest, Austria profoundly concerned over the future designs of both Russia and Serbia—had some quite direct connection with Sarajevo, no other conceivable outcome would have brought a guarantee of stability either.

Then, too, the Balkans were Europe's periphery, not its center. The happy citizen in Goethe's *Faust* had put it well:

> I know nought better on a Sunday or a holiday
> Than chat of war and warlike pother
> When off in Turkey, far away
> These people clash and fight with one another.
> We stand beside the window, drain our glasses
> And see how each gay vessel down the river passes
> Then in the evening homeward wend our ways
> Blessing with joy sweet peace and peaceful ways.

That was the way it still was. The world was not that interdependent. Balkan plots were subplots, and it would take an exceptional set of circumstances to change this.

What, though, of the conflicts between the major powers, whose story has just filled so many of these pages? What of colonial rivalries, of near-war over Morocco, of Strasbourg unforgotten, of dreadnoughts, of the adventures of Italy? Two things need to be said about these, if we are to put them into some kind of perspective. The first is that there was some consolation in their very seriousness. Europe's diplomats were quite as capable as later historians of seeing the dangers they faced, and since none of them desired general war, the incentive for finding some formula for peaceful compromise was considerable. (That such a formula was often agreed on only at the last possible moment was another matter, but then, the only new thing about brinkmanship is the term itself.) To quote Jules Cambon, in a report he sent from Berlin in January 1912:

> The failing health of the Emperor of Austria, the far-reaching plans attributed to the Heir-Apparent, the Tripoli War, the desire of the Italian government to extricate itself from the difficulties it had brought upon itself by mixing the disputes of others with its own, Bulgaria's ambitions, the threat of trouble in Macedonia, the difficulties in Persia, the shock to the credit of China, all pointed to serious disorders in the near future and the only hope was that the gravity of the danger might lead to its being averted.

The second point worth making is that while there were many things that went wrong in Europe between 1871 and 1914 (Cambon's list is only a sampling, after all), there were perhaps equally many that went right. Germany did restrain her ambitions for two decades. Britain and France showed that no matter how ancient their enmities and how present their imperial rivalries, it was perfectly possible to find a formula for accommodation. Germany and France reached no similar agreement, yet talks between the two never ceased. Strong forces in both countries worked toward reducing tensions, and by 1914, they had reached an understanding concerning spheres of economic interest, while the President of the Republic broke a habit of hostility of over four decades and came to dine at the German Embassy in Paris. Great Britain and Germany, real though their naval and colonial rivalries were, similarly continued to explore the possibilities of a *détente*, whose concrete results, in 1914, were an agreement to construct new battleships at a rate of 16 to 10 in England's favor, and an understanding under which the two nations agreed again to divide Portugal's African possessions to the exclusion of Europe's other powers. Russia did act to curb Belgrade in

1908. The Austrian government, for its part, did dismiss the chief of the general staff in Vienna when he advocated preventive war against Serbia.

In fact, the concept of war itself as a rational and legitimate means of policy was being questioned. The Hague Convention of 1899 not only initiated some serious international discussions on the limitation of armaments, it also established a Permanent Court of Arbitration, which first convened in 1901, and by its existence offered a juridical alternative to force. The list of what went right could be extended. But it may be enough to say that it was largely in deference to the conventions of historiography, which tend to deal with crisis rather than normalcy, and to stress disaster over triumph, that these pages have had so much more to say about Morocco than the Hague.

But even if we stress the confrontations, and downplay the accommodations, it is apparent that all of the crises mentioned were capable either of being contained, or of being kept from spilling over into general hostilities. To understand, then, the causes of the First World War, we will need to do something else, and survey at least two other topics. One of them will deal with the basic, underlying issues that made for conflict between the powers. The other will concern the immediate crisis that led to a state of war in 1914.

# CHAPTER 3

# Europe on the Eve: How Deep the Trouble?

According to one report (there was no Congressional Record then, and we rely on rumor) a member of Poland's parliament, the *Sejm*, once rose to stop the flow of an afternoon's particularly euphoric rhetoric. "Gentlemen," he said, "if everything is so good why is everything so bad?" If so much was right with pre-1914 Europe, how could a second-rate Balkan plot set off a first-rate catastrophe? Were there not, in truth, organic weaknesses that plagued Europe and lowered the continent's powers of resistance to a perilous degree?

What were the troubles, deeper than the immediate crisis, that affected Europe before 1914? We might do well to have at least a look at the principal diagnoses made by various writers: at the state system, at the system of alliances, at major sources of conflict both overseas and closer to the center, at the role of public opinion and the press, and finally, at the acceptance of war as a means of policy.

## A Matter of Political Organization: The National State

Some writers have suggested that one truly basic factor that made war an ever-present prospect in 1914, (as in 1814, as well as on the day that this is being read) was nothing less than the national state—a form of political organization accepted by a majority of the world's citizens as one of the immutable facts of life, no more subject to debate than, say, the sun's rising in the east. Actually, the argument for this point of view suggests, the national state was a relatively novel institution, replacing a feudal system in which differences of class, faith, guild, and estate had, on occasion, mattered far more than national dividing lines. The dangers already inherent in the sovereign state, which assumed most of its present characteristics in the age of absolutism of the seventeenth century, increased in the eighteenth and nineteenth centuries. For it was then that the people, as a whole, came of age politically, and the state was identified not primarily with its monarch but with the entire body politic. It was then, too, that nationalism

*German misgivings about Britain are suggested by this 1898* Simplicissimus *cartoon by Th. Th. Heine. The heading reads, "Anglo-German Treaties," and the caption, "Come on, Michael, give me your shirt, too—I'd like to see you completely naked." (Michael traditionally served as the symbol for Germany, as John Bull did for Britain or Uncle Sam for the United States.)*

tended to turn into a faith, replacing previous emotional, and sometimes formal, allegiances now dead.

An aspect of things that made this development especially explosive was that the limited kind of warfare that had characterized past ages was becoming a nostalgic memory—partly because of the advances of technology, but in even larger part, because it was no longer a case of Florence fighting Venice or of Tours having it out with Angoulême. Instead, it was one nation against another: all of France against all of Italy, with the concomitant release of hatreds and energies that had no equivalent in either the civic rivalries of the middle ages and the Renaissance, or in the comparatively civilized cabinet wars of the age of absolutism. National passions, organized or spontaneous, had entered the scene.

Even where these could be contained, something else that was very basic made for great peril. It was the permanent possibility of conflict between nations whose sovereignty was unrestricted. The sovereign state, that law unto itself, faced other sovereign states, and was, considering the propensities of independent entities, as likely as not to come into conflict with them. Yet there existed no effective organization that could intercede if violence should result from the pursuit of conflicting national interests. There was no higher, supranational authority that could be relied upon to arbitrate between nations bent on mutual destruction.

Where the beginnings of such an authority did exist, its powers were too circumscribed to be effective. The International Court of Arbitration at the Hague could act only if each one of the nations involved in some specific argument voluntarily agreed to submit the case to the Court's jurisdiction, and even then, the Court had no police powers with which to enforce its ruling. (This, incidentally, is a situation that continues unchanged to our day. The Court has changed its name, and been made a part of the United Nations, but its authority is precisely as limited as it was in the years preceding Sarajevo.) What Europe offered, in the phrase of G. Lowes Dickinson, British Liberal and pacifist, was the spectacle of "international anarchy." The national state, unchecked as it was in the exercise of its sovereign power, except for self-restraint born of wisdom or fear, was set on the road to disaster. Trouble, if we accept this interpretation of events, lay deep indeed.

## Poisonous Medicine? The Alliance System

What made the risks involved in international affairs even more terrifying was the system of alliances that had evolved by 1914. To a majority of powers, the answer to the international anarchy lay not in devising means to diminish national sovereignty, but in finding as many allies as they possibly could among the existing nation states. Faced as they felt they were by "reality," adding to their diplomatic strength by means of casting the widest conceivable net of alliances seemed very much preferable to waiting for Utopia. But should not Europe's statesmen have remembered Molière's words, from *Le Malade Imaginaire*, that "most men die of their medicines, not of their diseases"? Three major, and quite demonstrable, reasons would have seemed to indicate that in this case at least, the cure was worse than the disease.

One was that Europe, after 1907 at the latest, was divided into two hostile camps: the Triple Alliance of Germany, Austria-Hungary, and Italy versus the Triple *Entente* of France, Russia, and Great Britain. Europe's six, then, had chosen sides years before any actual outbreak of hostilities, and in the process had decreased, and perhaps even destroyed, the space for diplomatic maneuvering that was (and is) essential for the preservation of peace.

The second and related point was that under the alliance system, as it then stood, minor powers were in a position to make very major decisions. Some rash action on the part of little Serbia could involve giant Russia in the consequences. Russia, in turn, might involve France. Or Austria-Hungary, with its entanglements in Southeastern Europe, could compel Germany to go to war over issues that were of no imaginable, or certainly of no direct, concern to Berlin. In short, any crisis, no matter how trivial by any reasonable and objective standards, carried the seeds of catastrophe. Given the facts of the alliance system, an offense to any one European nation might ultimately draw in that nation's ally, and then that ally's ally, or allies. The tail could wag a whole pack of dogs.

The third charge, and it is still a related one, was that nations at times were tempted to play a more reckless diplomatic game than they would have under different circumstances. Had they stood alone, their relative weakness might have served them as a continual reminder of the need for compromise and caution, but as it was, they could feel confident that their alliances would shield them from the worst results of their actions. Thus Russian ambitions in the Balkans, for instance, had been rather effectively held in check during an earlier period by the opposition of a number of other powers. But was it not natural that Russia, after the conclusion of the accords with Britain and France, should be inclined to pursue a much less circumspect policy, since in the event of a collision with Austria, or even with Austria plus Germany,

St. Petersburg could count on the support of London and Paris? Thus Austria, for years, had been reluctant to pursue too irreconcilable a policy vis-à-vis Serbia. But, in 1909, as the Austro-German alliance hardened, the chief of the German general staff let his Austrian colleague know that if Austria should ever be compelled to march on Belgrade, and if Russia in such an event should come to Serbia's aid, Germany would be certain to go to war against Russia. It was a promise that Bismarck, with his far more skeptical view of alliances had never been willing to make; the idea that Austria might be encouraged to provoke Russia had terrified him. But now that Bismarck was dead and Austria could count on Germany's unquestioned aid, was it really very astonishing that she would act with such intransigence toward Belgrade in 1914?

This then, if we follow that particular argument, was the mood which the alliance system brought to Europe. And it brought it to a continent that was rent by national rivalries which had few precedents in past history.

## Conflict Overseas: Imperialism

One such rivalry, or rather series of rivalries, took place overseas. The two-score years before Sarajevo were years of all but unparalleled imperialist activity. Not that imperialism was a new fact of European life. The practice, if not the term, had been been quite as familiar to Ferdinand and Isabella as to Benjamin Disraeli. But imperialism seemed to have run its course. During the first three quarters of the nineteenth century, the powers of Europe were showing relatively little interest in acquiring any new territory beyond the confines of the continent. There was a good deal of economic imperialism, particularly on the part of Britain and France—in Latin America, in North Africa, in the Near East—but the outright acquisition of colonies tended to be regarded as an annoyance rather than an asset. The secession of the American colonies had been a case study to bear in mind.

In the 1870s, however, the attitude changed and changed abruptly. The powers were once again moving into Africa and Asia, and doing so with such energy and speed that by the time of Italy's Tripolitanian venture, the race was all but over. The end of Europe's overseas frontier had been reached. In the process, there had been acquired vast tracts of land and millions of new subjects. Africa, still largely uncharted in the 1880s, a generation later not only had some excellent maps, but roads and buildings, and hosts of white settlers as well.

The leader in the race was England, although other powers followed very closely. Nevertheless, it was the British who showed just how effectively one of the oldest of colonial powers could meet new opportunities, as they expanded into India, China, Persia, Egypt, the southern part of Africa, until by 1914, one fifth of the world's land and one fourth of the world's population were under the rule of Britain—the world's largest Empire since Imperial Rome. Then there was France, her gains particularly impressive for a country that, in the wars of the eighteenth century, had lost most of the vast colonial holdings accumulated in the period between Henri IV and Louis XIV. The Third Republic now did as well as any Bourbon monarch ever had, establishing protectorates over Tunisia and Morocco and colonies in Western and Equatorial Africa, and taking control of Indo-China and Madagascar, of Tahiti and a number of other similarly tempting Pacific islands. It was a vast empire, and one whose population easily outnumbered that of the mother-country.

Then there were the powers new, or relatively new, to the colonial game: Germany, Italy, Belgium. In Germany's case, Bismarck's restraint was wearing off. Colonies, he had said, reminded him of the splendid furs some impoverished Polish noblemen wore, with no shirt underneath. His successors felt differently about the matter. What they wanted were furs, not metaphors, and in their search they moved into East and Southwest Africa, farther north into Togoland and the Cameroons, and onto some Pacific islands as well. Italy, too, whose territory had only recently provided the scene for other nations' expansionist ambitions, turned from pursued into pursuer. The flag of Italy went up over Eritrea and Somalia, although an attempt to conquer the country bordering both, Abyssinia, ended in failure. At the battle of Adowa in 1896, the Abyssinians supplied the exception that proved the rule and startled the world with the news that Africans had beaten Europeans. But the Italians, as we saw, were to find another and happily feebler opponent in Turkey, and manage to incorporate Tripoli into their new empire. Belgium did even better as she acquired, by purchase, rather than by battle, Congolese territory seventy-five times the size of Belgium proper—territory, moreover, that was appreciably richer in resources than Tripoli.

Finally, there were the two powers that, while either non-European or partly European only, still were very much part of the imperialist

*Western Imperialism was never without its domestic opponents. Not least among the arguments against it was that it tended to brutalize the colonizers as well as the colonized, a point seldom made as wittily as in this 1904* Simplicissimus *cartoon by Ferdinand von Reznicek entitled "The Force of Habit."*

movement. One was the United States, which in this period annexed Hawaii, the Philippines, Cuba, Puerto Rico, and portions of Samoa, and made its presence felt on the Chinese mainland. The other power was Russia, whose imperialism largely followed the pre-1898 American pattern of expansion not so much overseas, as on her own continent. Like America's, Russia's aims were grand indeed. Russia was moving in the direction of the Middle East, or more specifically, toward Persia and Afghanistan, and toward the Far East, into Siberia, Korea, and the coastal regions of China.

In was, in sum, an unprecedented epoch of expansion. Of all the major powers, Austria-Hungary alone kept aloof, while Spain and Turkey, through lack of strength but not of desire, lost ground instead of gaining it.

What, however, was the connection between imperialism and war? The answer is that there were impressive arguments for maintaining that imperialism affected the policies of the European powers in several obvious, and ultimately disastrous, ways. There was, for instance, the new mood of violence that it helped to create. If ever one had daily proof of how might made right, imperialism provided it. Sweet were the uses of brute force: they had made the Queen of England Empress of India and created a new France beyond the seas. And it was violence that took place against the background of an often drab, dull, and tedious industrial society, thus making colonial wars seem all the more exciting and colorful.

As the great British historian and long-time Liberal member of Parliament, G. P. Gooch, wrote in 1902: "When we read on the bills: 'Boers sabred by moonlight,' we are supplied by a novel and striking image, which for a time relieves the monotony of life. The Romans clamored for '*panem et circenses*.' We have changed our religion and two thousand years have slipped by, but the cry is the same today." War, in an age where death was becoming mechanized as never before, was given an air of romance and glamor.

There was also another and even more concrete way in which imperialism could be said to have contributed to the destruction of the European order. It was that colonial ambitions required increased armaments. If one was to conquer the heathen—and be ready for possible trouble with some fellow missionaries as well—more money would have to be voted for arms, and so it was. The means as well as the spirit of violence were increasing. Very large amounts of effort went into the production of machineries of conquest. Very small amounts went into any real consideration of the consequences. In retrospect, we are startled by the recklessness with which the nations of Europe pursued their colonial dreams. In their greed, they appeared to be oblivious to the diplomatic repercussions of their Asian and African adventures. Thus, the 1880s and 1890s offered the almost constant spectacle of crises between England and Russia over who was to control Persia and Afghanistan. Thus, in the Pacific, Germany, England, and the United States would allow an inordinate amount of friction to develop between them over the precise delineation of interests on the island of Samoa; each nation's honor and survival, if one followed some of their respective spokesmen, were depending on the number of Polynesian palm trees it controlled. Thus, Germany and France, early in the new century, would twice go to the edge of war over the matter of spheres of influence in Morocco.

Lenin, looking on, thought it all very good news. Imperialism, he later wrote, was the final aberration of capitalism, the most fateful, perhaps, of capitalism's many inherent contradictions that would speed its end. How far from wrong was he, when we add to the sampling of near-wars the real wars that imperialism caused, from the Boer War to the Boxer Rebellion to the Russo-Japanese War of 1904–1905? Now it can be said that these wars were fairly remote in time and cause from World War I (an argument to which we will return later). But one aspect of imperialism that was not remote at all was the manner in which it poisoned relations between Germany and England. In the end, it would be the hostility between these two countries that would do as much as anything to transform a Balkan assassination into a world war.

## Conflict at Home

### *Dardanelles to Alsace*

Before dealing with the Anglo-German problem in any detail, however, it might be wise to cast a brief glance at Europe's basic trouble spots, *circa* 1914. For there existed some quite profound areas of friction between Russia and Turkey, between Russia and Austria, and between Germany and France, in addition to those between Germany and England. There existed others as well, of course; conflicts of interest are the daily bread of diplomacy. Yet it was on those four areas that the historians' attention has rightly concentrated. It was here that most of the explosives accumulated.

Russia's foreign policy, to start with the border region of Europe, was dominated, as it had been for two centuries, by the desire for access to the open seas and to the trade routes of the world. Specifically, this meant control of the Straits, and given the weakness of the Porte, the goal seemed worthier of pursuit than ever.

There was some irony in this, since by this time, the Straits were a matter of prestige rather than necessity to Russia. To quote George Kennan:

> The regime of the straits that existed in the period prior to the war occasioned no difficulties for Russian commerce. And if one looks at the matter from the military standpoint, one can only say that this region, barring as it did the passage of foreign warships through the straits, was a positive advantage to a country which had lost most of its naval strength at Tsushima.

Still, this *was* the aim Russia pursued, and it was a policy of great inherent risk. St. Petersburg was willing to take the risk because it had not quite adjusted to changed conditions; the habits of the past were more compelling than the realities of the present. Besides, even where the statesmen's vision might be sharp and clear, there was the problem of Pan-Slav pressure.

Pan-Slavism, like official Russia's desire for the Straits, does not truly make a great deal of sense when looked at objectively in the second half of the twentieth century. To quote George Kennan again: "Membership in the Slavic branch of the human family is not really, as history has shown, a very important bond. Cultural and religious traditions, varying widely among the Slavic peoples, are more important as determinants of national policy, in the long run."

Yet there it was. The actual number of dedicated Pan-Slavs in the Russian diplomatic service was relatively small; still, the sentiment and the pressures were there. No Russian foreign minister could entirely disregard them. And Pan-Slav pressures, combined with more traditional Russian foreign policy aims—the line dividing the two was a thin one in any event—placed a heavy burden not only on Russo-Turkish relations, but inevitably on Austro-Russian relations as well. The weakness of Turkey was an open invitation to both Russian and Austrian expansion in the Balkans, since none of the successor states to Turkey were viable enough to be wholly independent. True, statesmanship of a very superior order might have avoided or at least mitigated the resulting friction between the two great nations. If, on the one hand, the Austrians had acted with greater wisdom and restraint in 1908 and not faced the Russians with the choice between public humiliation and war, subsequent history might well have taken a happier course. So it might, on the other hand, if Russia had not involved herself quite so deeply in the affairs of Serbia, for what was at stake there was more than a competition for Balkan spoils. It was a question, rather, of national survival, since Belgrade, with Pan-Serb ideas of its own, posed a clear threat to the very existence of the Habsburg monarchy. But these are speculations. The fact was that both powers possessed foreign ministers of merely human ability and imagination and that the pre-1914 Balkans provided another perennial trouble spot where crisis seemed normal and stability utopian, and where major powers were willing to take some terrible risks in the pursuit of gains that, from our vantage point at least, seem wholly out of proportion to the price.

Then, closer to the center, there was the enmity that came between France and Germany. We can dispense with a long discussion of its roots; enough has been said about the conditions of the peace of Frankfurt. No more needs to be added here than to reiterate the fact that the French would not forget Alsace-Lorraine, and that a border region which should have been a bridge (and is one today) became a chasm. One of the reasons, for instance, for General Boulanger's vast popularity in the 1880s, and for his near success in destroying the Third Republic, was his supposed toughness toward the Germans:

> See him over there, he smiles as he goes by
> He has just set free Lorraine and all Alsace!

He had done nothing of the kind, nor did the duly constituted governments of France have the power to correct Frankfurt by diplomatic means. The Germans, for their part did not have the wisdom to do so, and so an unjust peace remained to divide two nations whose enmity was anything but eternal, and who in our time provide the bedrock of a continent united in the European community.

## *The Sources of Anglo-German Friction*

The Germans did not even possess enough good sense to avoid policies that were bound to create friction with a nation from which no victor's peace kept them apart, that is, with Great Britain.

The causes, on the German side, went beyond any of the specific points at issue. Germany, under William II, wished to be a great power.

Why not, one might say. For Germany, exuberant in her new-found unity, possessed the population, the military strength, the industrial capacity, the vital energy, that go into the making of a great power. Why not, then? Mainly because Bismarck was gone and the conduct of Germany's foreign affairs rested in hands less steady than his. There was a very great deal of talk about Germany's might, about how Germany had been forgotten when the goods of this earth were distributed, about how Germany must trade the pursuit of limited objectives for that of *Weltpolitik*. There was very little clear thought, on the other hand, of what precisely it was that Germany wanted. How really does one translate *Weltpolitik?* "Global policy" perhaps, a term as hazy, and at the same time as ominous, as the original. The translator's difficulties are symptomatic.

The world, not surprisingly, was puzzled and concerned about the limits of Germany's ambitions. What were the Germans pursuing? Was it colonies, was it a new eminence on the continent, was it, perhaps, world domination? We know today that most foreign fears were groundless, and that no responsible German statesman before 1914 thought in terms of Napoleonic visions. At the time, the matter was less clear, as the Kaiser's far from subtle attempts to make Germany the equal of Great Britain, and above all, his ill-considered speeches allowed the impression to flourish that his country was a menace to peace, security, and a decent society. Many a German could also see the danger signs. Theodore Fontane, Germany's greatest nineteenth-century novelist, wrote to a friend:

> You mention the speeches given by an exalted tongue, in which so much is said and even more is passed over in silence. I always lose my temper when I read them. . . . He has a million soldiers and wants to have a million battleships too. He dreams . . . of humiliating England. Germany is to be on top, in each and every thing. I rather like all this—I won't discuss right now whether it is clever or practical—and I would gladly follow him on his tightrope walk if only I could see the right sort of chalk under his feet and the proper balancing staff in his hands. But those are the things he does not have. What he wants, while it may not be impossible, is vastly dangerous, and his equipment is wrong and his means are insufficient.

Not all Germans were as cautious as Fontane. There had come into existence, simultaneously with William's "New Course," the small but vocal group of Pan-Germans. Their specific aims were even more ill defined and fluid than those of Germany's official spokesmen. They talked of fusing such "Germanic" nations as Holland and Switzerland with the Reich. They dreamt dreams of incorporating, by political or economic union, regions reaching from Brussels to Bucharest into a German-dominated *Mitteleuropa*. They conjured up visions of a truly grand empire in Africa. The precise aims might change; the spirit remained the same. The Pan-Germans forever favored an aggressive and acquisitive approach to foreign policy; they were forever ready to engage in more tightrope walking, secure as they were in the knowledge that there were no ropes like German ropes and no nets like German nets.

The Pan-Germans were a minor group. They were less influential than their Pan-Slav counterparts in Russia, which was fair enough since their program was even more at variance with the world's realities. The largest membership they ever claimed at any given point was well under twenty-five thousand, and for the most part it was considerably smaller than that. Yet there existed several factors that made it difficult to deal with them as a mere fringe group. One was that among the members were a fair number of Reichstag deputies; in 1901, their most successful year, thirty-two Reichstag members were Pan-Germans. Another was the volume of the noise they made. A third was that the Kaiser's government, while not making Pan-German aims its own, failed to disassociate itself from their inanities with sufficient force. A fourth was that there were people beyond Germany's frontiers who were quite happy to make sure that the Pan-Germans received a wide hearing.

This applied, in particular, to Great Britain, where any indication of the existence of a German menace found some willing listeners. "It is of course true," noted a senior member of the British foreign office after reading a long dispatch on Pan-German activities, and making quite sure that it would reach the king, "that the Pan-German aspirations to dominion over the Low Countries and over the Adriatic are openly disavowed by all responsible people in Germany. But it would be foolish to doubt that if and when a favourable opportunity occurred for realizing such political aspirations in whole or part, the opportunity would be seized by the German Government with all its wonted energy."

The causes of this sort of suspicion, and of Anglo-German friction in general, were not all on one side of the channel, and on the British side, too, they transcended the individual points at issue. There were, on that side, factors that may make one appreciate why people as sensible and civilized and fond of many things British as Fontane was, wished the Kaiser luck in his policy toward England.

There was, for one thing, an amount of arrogance which exasperated people other than the Germans, too. God, to listen to some British speakers, had left the side of the bigger battalions where He had been in Frederick the Great's days, and permanently joined His cause with Britain's. There was, behind the arrogance, national expansion and power that had few parallels in history. It was, very likely, too much power, since the British Empire covered so much of the globe that just about any move on the part of just about any nation just about anywhere in the world could be considered a menace to either some part of the Empire or to its lines of communication. And the principal challenge to the Empire seemed to come from a nation too newly rich to abide by the rules laid down by British gentlemen, from the Kaiser's Germany. It was a threat to be resisted for moral as well as political reasons; one is back with the matter of arrogance. To quote the distinguished American diplomatic historian Raymond Sontag:

> Nearly all men, and groups of men, assume that what is good for them is good for the world, and that what is bad for them is bad for the world. Englishmen forgot their history and assumed that their supremacy resulted from the operation of a beneficent moral law. To challenge British supremacy was to challenge moral law. Could the Germans really be so wicked? Roughly speaking, the Conservatives believed that the Germans were capable of this crime; the Liberals were reluctant to entertain such a thought.

That the Conservatives believed it might not have been too tragic, since it was two Liberal ministries, first that of Campbell-Bannerman and then that of Asquith, that held office after 1905. But among those who believed it were some of the permanent officials of the British foreign office as well. A passage written by one of them may serve to indicate their views; it may also illustrate the British attitudes just described. It comes from a memorandum written in 1907, by Sir Eyre Crowe, then the Senior Clerk of the Foreign Office. In a survey of Britain's relations with Germany and France, he had this to say as he came to discuss the matter of naval armaments:

> The danger of a general combination against Britain's naval supremacy can in practice only be averted on condition that the national policy of the insular and naval state is so directed as to harmonize with the general desires and ideals common to all mankind. . . . Now the first interest of all countries is the preservation of national independence. It follows that England, more than any other non-insular Power, has a direct and positive

> interest in the maintenance of the independence of nations, and therefore must be the natural enemy of any country threatening the independence of the others, and the natural protector of the weaker communities. . . . It has been well said that every country, if it had the option, would, of course, prefer for itself the power of supremacy at sea, but that, this choice being excluded, it would rather see England hold that position than any other state.

But that happy state of affairs was in jeopardy, and the source of the threat was plain to Crowe:

> A German maritime supremacy must be acknowledged to be incompatible with the existence of the British Empire, and even if that Empire disappeared, the union of the greatest military with the greatest naval power in one State would compel the world to combine for the riddance of such an incubus.[1]

*Le style c'est l'homme*. England, "the natural protector of the weaker communities." England, the chosen master of the seas. And why? Because "it has been well said" that this should be so. Because the possibility of any other nation's being supreme at sea was "being excluded." But by whom had it "been said"? By whom was it "being excluded"? How simple the problems of the world become in the passive voice. German naval supremacy? (A term which, when properly translated, should read "Germany's increased naval strength," since supremacy, even if the Germans had aimed at it, which they did not, would have been decades away.) It "must be acknowledged to be incompatible with the existence of the British Empire"—after all, that Empire was Germany's "natural enemy."

It was the voice of neither precision nor reason. When practicing diplomats begin to think of other powers in terms of natural enemies and of incubi, rather than in terms of nations with aims, interests, needs, and desires of their own—some legitimate, some to be resisted—the end of meaningful discourse is in sight. Perhaps the one thing that "must be acknowledged" is that Germany was not the only country to miss a grand old man. Where was Disraeli? It is hard to visualize him tolerating a memorandum such as this. Sir Edward Grey, in whose hands the conduct of British foreign affairs now lay, not only tolerated it,

[1]Sir Eyre Crowe, *British Documents on the Origins of the War, 1898–1914* (London: H.M. Stationery Office, 1928), III, Appendix A.

but recommended it to the attention of the prime minister. Crowe's observations, he wrote, were "most valuable. The review of the present situation is both interesting and suggestive, and the connected account of the diplomatic incidents of past years is most helpful as a guide to policy." Grey was a man of great personal probity and attraction, yet, despite his long tenure of office (he served as under secretary for foreign affairs from 1892 to 1895, and as foreign secretary from 1905 to 1916) there was about him something of the air of the gifted amateur. Or, if that is too unkind a term, he strikes one as the kind of officeholder under whom expert advisers become makers of policy, and who, rather than shaping events, is all too easily controlled by circumstances.

Men who shape history are admittedly rare. Yet in the absence of genius, on the British as well as on the German side, the Anglo-German dispute continued unresolved. At least three specific problems and a matter of attitude combined to separate the two countries.

There was, in the first place, the economic rivalry between the two powers. The problem was not limited to England and Germany. Trade rivalries existed between a great many European states, and some writers have seen them as one of the truly basic problems aggravating European international relations before 1914. But Anglo-German competition probably was the sharpest of them all. German industrial growth, both in absolute terms and vis-à-vis England, had been phenomenal. In 1870, for instance, Germany had produced 169,000 metric tons of steel as against England's 286,000. In 1910, the figures read 13,698,000 for Germany and 6,374,000 for England. Not all figures were this spectacular; nevertheless, in coal production, Germany increased the yield by 800 percent between 1870 and 1914, England by 200 percent. In similar fashion, the Germans showed by their output of finished goods, particularly of chemicals and of electrical equipment, that while the industrial revolution had come late to their country, it had found a congenial home.

Statistics on exports were keeping pace with those on production. In 1871, for instance, German exports to the United States were well behind those of Great Britain; in 1898, they had nearly caught up; in 1913, they were nearly three times as high. In terms of total exports, Germany, while starting as a poor second indeed, by 1914 had become all but the equal of Britain. It was perfectly true that the world offered enough markets for both these industrial giants, yet the daily competition for sales tended to dramatize the conflict that existed in other fields as well. How was a British member of parliament to react when, during a foreign policy debate, he was to glance at his pencil and see the legend "Made in Germany"?

With trade rivalry went another problem, that of colonial friction. Too often, when the British wished to expand, whether it was in Africa or in the Pacific, there—or so at least it seemed to London—were the Germans, asking for "compensation." Nor would the British very easily forget the gratuitous insult of the Kruger dispatch. With the Germans in Southwest Africa now and in the Cameroons and in East Africa, unbidden and unwanted neighbors to Britain's domains, the memory was likely to remain fresh.

Why this German hunger for colonies, why the break with Bismarckian tradition? In part, German imperialism makes sense, even today. An expanding German industry was in need of markets and, to an even larger degree, in need of raw materials. Yet there is no absolute need to push forward the frontiers of empire in order to assure oneself of a proper supply of kapok from the Cameroons or of cloves from Zanzibar. Less aggressive ways of commerce may work equally well; certainly the Swiss have grown extraordinarily rich without acquiring a single colony. Switzerland, it can be replied, was a small power, Germany a large one, which could not afford to rely on the good will of her potential enemies for the materials needed to sustain the nation's economy. Still, even when such supposedly strategic economic needs are taken into account, there remains an element in Germany's colonial appetite that defies rational explanation. For economic blockades in time of peace tend to be as rare as they are ineffectual; nor was this contingency one that necessarily entered the minds of Germany's leaders. Trade flowed too freely in the years before 1914. It was rather that some Germans were getting intoxicated with their own slogans—of Germany's might, of keeping up with Great Britain, of needing a "place in the sun." Here, as elsewhere, there was much action for action's sake, and little thought about what the limits of Germany's ambitions should be.

Much of this also held true of the third issue that came between Germany and England, the naval race. "Our future," to recall the Kaiser's elegant phrase, "lies on the water." Deeds had followed his words. Under the leadership of Admiral von Tirpitz, Germany's all too energetic Secretary of the Navy from 1897 to 1916, the Germans proceeded to build a navy that was meant to be second only to that of Great Britain.

Some of the motives behind Tirpitz's program were rational enough. Both the expanding German Merchant Marine and the colonies that now flew the German flag deserved the protection of a German fleet. There may also have been an element of sense in the admiral's idea of the "risk navy." Let the German navy, argued Tirtpitz, not aim at supremacy. That was a condition plainly beyond the country's potential. Let Germany's navy, however, be strong enough so that no other and superior naval power would think of attacking it for fear of the losses that such an attack might entail. Let it, in modern parlance, be a deterrent navy. Faced with such a navy, and here came the second, rather less solid part of Tirpitz's argument, the British would surely persuade themselves of the wisdom of reaching some political agreement with Germany.

This, as we know, was a miscalculation. It is not very often that nations can frighten each other into friendship, and the British reaction, not very surprisingly, was that of the clenched fist, and not of the outstretched hand. Not very surprisingly, for what did the Germans really have in mind when they embarked on Tirpitz's program; to what extent had they considered the consequences?

Historically, Germany was a land power with hardly any naval tradition. Looking at the new German battleships coming off the yards, one is struck by how many of them were named after Prussian and German generals—Derfflinger, Moltke, Seydlitz, Goeben, Blücher—since there were simply not enough admirals to lend their names to the christenings. Obviously the British would be vitally concerned, even with people more balanced than Crowe in the foreign office, at the spectacle of Germany, the possessor of Europe's strongest army, setting about the creation of a major navy as well. The direct challenge to Britain's control of the seas quite aside, it was too much power for any one country to hold.

Why, then, did the Germans go ahead, plain though the unfortunate effects of Tirpitz's ambitions on Anglo-German relations were? Why did the Kaiser continue to give the admiral his blessings; how could he tell his Chancellor, as late as the fall of 1913, that "England will come to us not in spite, but because of my Imperial Navy"? There was operative here, one suspects, the same element that made the Germans proceed with equal disregard for caution in the colonial sphere: a burning desire to be recognized as Great Britain's equal, and with it, a suspicion that the desire was not likely ever to be fulfilled.

Out of place as the phrase may appear in a diplomatic narrative, there was a deep psychological bond as well as bar between Britain and Germany. What came into play here was a curious love-hate relationship, aggravated by some profound feelings of inferiority on the German side. It could be found in William II, who delighted in seeing himself in a British admiral's uniform, who was proud of being half-English (his mother had been Queen Victoria's eldest daughter), who loved reading English books, and who spoke and wrote English with facility and pleasure, and who, at the same time, was worrying whether the British were taking him quite seriously, and behind whose nearly every move lurked fear of being secretly smiled at, of not being truly considered the equal of his cousin Edward and of Eton's best. It could be found among many educated Germans who sensed that no matter how much of Britain's tailoring, country clubs, parliamentary institutions, or orange marmalade they might imitate or import, they would never quite catch up with their cousins beyond the Channel; that they were newly rich but not yet newly wise; that they might be able to buy all the Harris tweed they wished, but that with all the cloth of the Hebrides, they would still lack the proper school tie.

Here is a passage from a German novel, *Peter Moor's Trip to the Southwest,* by Gustav Frenssen, that may underline some of this. The book appeared in 1906 and was one of the year's best sellers. Its narrator and title hero is a good and simple fellow who has enlisted in the German marines and at the book's beginning is on his way to German Southwest Africa to fight in a campaign against a native rebellion. As his boat passes the Cliffs of Dover, he chances to overhear the following conversation between an officer in the medical corps and a deck lieutenant:

> I heard the lieutenant say: "We sailors have different ideas about the British. . . . We meet them in all the world's ports and we know that they are the most reputable people of them all. There, behind those chalk cliffs, lives the first nation on earth: noble, wise in the ways of the world, brave, united, and rich. But we? For years past, we have had just one of their qualities, bravery. Slowly now, we're gaining one more, wealth. But will we ever gain the rest? There now is the vital question." The words surprised me. But later, the old Africa hands I met also spoke with great respect about the British.[2]

[2]Gustav Frenssen, *Peter Moor's Fahrt nach Südwest* (G. Grote: Berlin, 1906), pp. 14–15.

7. Jahrgang

Preis 15 Pfg.

Nummer 15

SIMPLICISSIMUS

Illustrierte Wochenschrift

Deutsche im Ausland

*While German suspicions about Britain could hardly be denied, neither could German admiration for things British, as suggested by the Frenssen quotation in the text, or by this July 1902* Simplicissimus *cartoon by Th. Th. Heine. The heading reads, "Germans Abroad," and the caption on the left is "That is the way Herr and Frau Schmidt looked when they traveled to London to watch the Coronation," and the caption, right, "And this is how they looked when eight days later they returned as Mr. and Mrs. Smith."*

Too often, the respect was not being returned. The "you are to love me!" approach is rarely effective. The Germans might have been well advised to drop the matter. How many foreigners, after all, did the British either love or respect? The Germans rarely paused to consider that the answer most likely was "none." An occasional exception was the Kaiser. "He is a thorough Englishman," he told Theodore Roosevelt when George V had succeeded Edward VII in 1910, "and hates all foreigners but I do not mind that as long as he does not hate Germans more than other foreigners." The insight did not last. Affection, as it so easily can, turned into hatred. Nor did the ethnic kinship that existed between the two nations stay the process. It certainly never acted as a substitute for missed alliance opportunities. On the contrary, it had a way of making things worse. Few feuds are as bitter as family feuds. In the end, when war came in 1914, one of the most wildly popular poems of the hour would be a "Song of Hate Against England," by a young and hitherto obscure German poet, Ernst Lissauer. Whom did the Germans really wish to see humbled? he asked. The French, whom they had battled in so many wars? No, not the French. "We love them not, we hate them not." The Russians, then, the ancient Slav antagonist of the Germans? No again. "We love them not, we hate them not"

> What's the Frenchman to us, or the Russian foe
> We'll meet shot with shot, and blow with blow.

No—"We all have but one foe, England":

> Hate by water and hate by land
> Hate of the head and hate of the hand
> Hate of the hammer and hate of the crown
> Hate of seventy million choking down
> They love as one, they hate as one
> They all have but one foe: England

## A Matter of Publicity: The Press

The topic of Anglo-German friction has been stressed at such length because in a sense it was the ancient nursery rhyme's horseshoe nail. Without it, there might have been no *Entente Cordiale;* without that *Entente*, French support for Russia might have been less resolute; without that support, in turn, Russia might well have acted as a restraining influence on Serbia. For lack of an Anglo-German *détente*, in short, more than one kingdom would be lost.

What aggravated the points at issue between Germany and England even more was the publicity they received. The Germans not only had a navy, they also had a Navy League, and so did the British, and both these groups sponsored more speeches, books, pamphlets, and letters to the editor than was beneficial to the cause of peace. Nor was there any need to rely on special interest groups to worsen, or at times to create, international disputes; the press, with no particular prodding, did quite well enough at that.

The end of the nineteenth century had seen a new mass literacy, and with it, the creation of a mass-circulation press. The managers of that press were quick to learn that bad news sold more copies than good. "They have ceased to publish the 'Newgate Calendar' and 'The Pirate's Own Book,' " wrote Ralph Waldo Emerson, "since the family newspapers . . . have quite superseded them in the freshness as well as the horror of their records of crime." Emerson was angry, but was he exaggerating that much? "Boers sabred by moonlight." William Randolph Hearst was not the only magnate of the press to discover that crisis and war—stimulated, or even, with great ingenuity, created *ab ovo*—made great circulation builders. Many a European counterpart to Mr. Hearst would emulate him, exaggerating minor incidents into major crises, presenting another nation's new arms budget as a threat to his readers' survival, or converting the possession of some particular stretch of African jungle into a matter of national honor. The press, in sum, played a substantial part in accentuating international friction, and perhaps even in preparing a mood for war. To quote the noted American historian of the period, Sidney B. Fay:

> Too often newspapers in all lands were inclined to inflame nationalistic feelings, misrepresent the situation in foreign countries, and suppress factors in favor of peace. . . . There is a vast literature on freedom of the press, censorship of the press, slander and libel, and the professional aspects of journalism, but there is very little sound writing on the relations of the press to governmental control and on its influence in fomenting national hatreds and war.[3]

The role of the press may be an aspect of Europe's pre-1914 troubles that stands in least need of further elaboration and illustration, even

---

[3]Sidney B. Fay, *The Origins of the World War* (Crowell-Collier and Macmillan: New York, 1930), I, 47–48.

though some of the volumes Fay had in mind still remain to be written. We are all quite familiar today with seeing those foreign statesmen whose policies or interests, at some given moment, run counter to ours, pictured as imbeciles, thieves, or threats to world peace in our own media. Still, we may draw back a bit as we come across some of the offerings of the pre-war press.

There was the series of reports on Germany in London's mass-circulation *Daily Mail* in 1909, for instance, whose resonance was strong enough to lead to their being reprinted in pamphlet form. "Germany," the reader learned, "is deliberately preparing to destroy the British Empire." Nor was that the extent of Germany's ambition. "All of Europe is to be Teutonised. We are all to be drilled and schooled and uniformed and taxed by Prussian officials, and the Emperor William II is to rule us with a rod of iron." Britain could respond with nothing less than the creation of a larger navy and the introduction of compulsory military service. "Britain alone stands in the way of Germany's realization of world-power and domination."

The attitude was not restricted to that part of the press called "yellow" (as though that epithet were enough to discount its influence). Some years earlier, London's *Saturday Review*, a journal much smaller in circulation and normally far more literate in tone, had carried an article that suggested that in

> all parts of the earth, in every pursuit, in commerce, in manufacturing, in exploiting other races, the English and Germans jostle each other. . . . Were every German to be wiped out tomorrow, there is no English trade, no English pursuit that would not immediately expand. Were every Englishman to be wiped out to-morrow, the Germans would gain in proportion. Here is the first great racial struggle of the future: here are two growing nations pressing against each other, man to man all over the world. One or the other has to go; one or the other will go.

The proposed solution was even more simple and radical than that of the *Daily Mail*. Destroy the Germans! "*Germania esse delenda*."

For once, it is easy to sympathize with the Kaiser, who after reading the *Daily Mail* articles said, "They are all mad in England, and people seem to think that I am standing here with my battle axe behind my back ready to fall upon them at any moment." His point might have been even stronger had the German press, in its entirety, been that much more restrained. Instead, some of the notes that *Saturday Review* and *Daily Mail* had sounded were being returned from across the channel.

Nor were a number of papers elsewhere on the continent reluctant to join in. In all, there are few more succinct comments on all this public venom than Humbert Wolfe's lines,

> You cannot hope to bribe or twist,
> Thank God! the British journalist.
> But seeing what the man will do
> Unbribed, there's no occasion to.

## Arms and the Men: War as a Means of Policy

"Power without responsibility," Stanley Baldwin is reported to have said apropos of the press, "the privilege of the harlot throughout the centuries." The words were fairly mild if one considers the damage the press was capable of inflicting in an age in which war was still an acceptable means of policy, yet in which the physical nature of warfare was changing beyond the imagination of man.

Here lay another very basic trouble spot of pre-1914 Europe. No matter what was happening in the laboratories and on the military proving grounds, war, by implicit agreement among the nations remained the *ultima ratio regum*, the final argument of kings, the ultimate and permissible resort in disputes between sovereign states. In such an atmosphere wrote the German historian, Golo Mann,

> war, as a concept and an institution, is an important factor contributing to war itself. Why should not something assume real shape from time to time which in one's thoughts one considers quite normal, which forms the high point in the life of nations, which is the standard by which the affairs of state—in their ranks, values, and morals—tend to be measured.

Yet how much better off Europe would have been to search for other standards by which to measure the affairs of state. For what made the unbroken acceptance of Clausewitz's celebrated dictum of war as the continuation of policy by other means especially ominous was the changing nature of those other means. Weapons were becoming increasingly destructive, and there were too many of them. The years that reached from the Franco-Prussian War to Sarajevo were the years of the great arms race between the powers of Europe. Increased expenditures for armaments were motivated in part by steadily escalating technological improvements, in part by the desire to back conflicting national ambitions with sufficient force, in part by the need to keep each nation's military preparedness on a par with that of its potential enemies. No matter what the motives, the speed with which the race was run had few precedents. Here, then, are some of the basic statistics on the growth of European armaments in the period, expressed in terms of per capita expenditures as translated into dollars:

| | *1870* | *1914* |
|---|---|---|
| Great Britain | $3.54 | $8.23 |
| France | 2.92 | 7.07 |
| Russia | 1.28 | 3.44 |
| Italy | 1.38 | 3.16 |
| Germany | 1.28 | 8.19 |
| Austria-Hungary | 1.08 | 3.10 |

Although Great Britain continued to lead in absolute terms, by far the greatest single increase was made by Imperial Germany, and a more detailed breakdown will indicate that the rate of German armaments accelerated with particular speed in the years immediately preceding 1914. Yet, while the specific rate of increase might differ, there was no country which felt that in the pursuit of security and power, it could keep out of the arms race altogether. The irony was that power and security are elusive things, and in the end, no country felt much safer than before, since the *proportionate* distribution of arms remained very much what it had been to begin with. There was, on the contrary, good reason to feel less safe, since the line between defensive armaments and the readiness for aggression was all but impossible to define.

The nations might have tried to halt the trend had they been able to visualize the nature of a war fought with the weapons whose deadliness was being improved with such ingenuity and skill. But most people thought of war in romantic terms, in terms of gallantry, devotion and comradeship, all qualities that would turn out to have comparatively little meaning in the face of the mass-produced slaughter for which the world was in fact preparing. A few of the experts knew better, but even they were wrong; for they thought in terms of quick, decisive campaigns and thus evaded contemplating the full discrepancy between the aims of twentieth-century warfare, on the one side, and its means, on the other. What no one foresaw was the kind of murderous yet indecisive warfare that would open in 1914. The generals played their war games and speculated, often with considerable intelligence, about the forms of future warfare, but not one among them envisaged what would take place between 1914 and 1918.

So the war games went on, and the statesmen did little to curb them. Here is a final aspect of the question "How Deep the Trouble?" that should at least be mentioned. Where were the inspired statesmen of early twentieth-century Europe? Where were the men of power whose overriding desire was peace, and whose abilities matched their motives? They were dead. The nineteenth century had known them—Talleyrand and Metternich, Bismarck and Disraeli. The early twentieth century could only honor their memory.

## But How Deep the Trouble? Some Conclusions

A formidable indictment, it seems. The national state and the alliance system, conflict overseas and conflict at home, a growing arsenal of arms and a press out for blood, an acceptance of war and a paucity of statesmanship. Was not a continent that permitted all these conditions to exist doomed in any event? At first glance, yes. At a second glance, we may begin to wonder; and at a third, the doubts will deepen.

For each and every point of danger mentioned, while real enough, had quite another side to it, too. The state system? But that was hardly new, and there neither was nor is any law stating that national sovereignty must automatically lead to war. In effect, supranational forms of government have either been very rare—there was but one *Imperium Romanum*, and even in its case, an occasional fight was necessary to maintain peace—or they have been illusory, such as the Holy Roman Empire, which in sober truth abounded with war. The national state, in short, was a reality and had been one for some time. Such a state, in the absence of a higher authority, could go to war; but it was equally free to remain at peace, and statesmanship, common sense, self-interest or fear were all powerful factors pointing to the desirability of the latter course.

The alliance system, then? Perhaps, if the alliances had actually been as binding and inflexible as some writers have made them out to be. But they were not. The British would consistently refuse to make a clear-cut promise to come to the assistance of France, let alone of Russia, in case of war. Every one of the alliances was quite specifically defensive in character, and as the Italians were to show, it was perfectly possible to change sides with impunity. In fact, instead of assailing a system that in essence was and is no more than the common pursuit of common interests (a difficult ambition to outlaw) might one not rather say that if only the alliances had been stronger, Europe would have been less troubled? To give but two specific illustrations, if for instance the Austrians had been certain beyond a doubt that an attack on Serbia would be an attack on Russia, or if the Germans had been altogether convinced that to invade France would, undoubtedly, involve them in war with England, might not history have taken a happier course? (A matter of speculation? Yes. But it is the basic assumption on which the western alliance system rested for years after the Second World War.)

What, though, of imperialism? That its effects on Anglo-German relations were deplorable should need no further elaboration. However, it should be said that it made Anglo-Russian and Anglo-French relations fully as difficult. There was far more colonial rivalry between Britain and France, for instance, than between France and Germany—the Germans and the French, after all, had no Egyptian campaign of 1798 to fight over again—and the Anglo-Russian wrangling that went on over Persia alone was every bit as sharp as any Anglo-German conflict of interests in Africa. Yet for all their imperialist rivalries, England, Russia, and France would ultimately find an alliance both possible and profitable. That imperialism could accentuate trouble was evident. What was equally evident, however, was that it could, as the case of the relationship between the United States and Great Britain after 1898 showed, bring about the realization that the interests of two nations were best served by cooperation. Beyond this, one ought to say that the absence of colonial ambitions was, in at least one instance, a cause for regret. To quote A. J. P. Taylor, the British historian whose insights could be as perceptive as his fallacies could be mischievous: "All the Great Powers except Austria-Hungary found a safe channel for their exuberance in expansion outside Europe. They stumbled on this solution by chance, without foresight." If only it had been possible to deflect Austria's interests from the Balkans to the South Seas, how much better that would have served the cause of European peace.

What of the more specific conflicts between the nations of Europe, the quarrels closer to home? Again, one hesitates to say that by themselves they doomed Europe in 1914. For surely one of the common denominators between the various issues that set Russia against Turkey, Austria against Russia, or France against Germany, was that none were worth a war. Russia might want the Straits, and want them badly, but given the opposition of Britain, France, and Austria, she was quite unwilling to risk an armed conflict over them. Austria might argue every inch of the way with Russia over Balkan spheres of influence, but the choice made was to let the diplomats and not the generals do the battling. France might seize every chance to change the verdict of Frankfurt, but it was every chance short of war. Whatever the period's equivalent of Henry IV's favorite city, it was worth a mass, but not the lives of one's countrymen. Or, to put the matter in another perspective, Russia, in the period that followed the First World War, faced territorial grievances infinitely worse than she, or for that matter, any other European nation confronted before 1914, *without* taking recourse to war to rectify them.

Besides, nothing is forever in diplomatic history. Russia and Austria, for instance, had lived through sustained periods of diplomatic harmony as well as of hostility; few powers had been closer partners than

*Real as the various sources of friction were in the years that led up to 1914, so was the desire for peace, and more than that, the idea that war was no longer an option between civilized nations. That mood is reflected in this 1908 cartoon from* L'Assiette au Beurre, *in which French troops face German troops, and the command "Ready" is followed by "Yes!" and "There!" with the soldiers on both sides turning on their bellicose leaders.*

St. Petersburg and Vienna during the quarter century or so after the Napoleonic settlement. Circumstances had changed, but so they might again. There was no reason why alignments should not begin to shift once more, why partnership could not replace rivalry. This applied with equal force to Anglo-German relations. Again, none of the specific instances of friction mentioned were precisely trifling in nature, but neither were they irreparable. Trade rivalry there was, but there was also beginning to be an understanding on the peaceable division of the world's markets, with the British concentrating on trade with the Empire, the Germans on that with continental Europe. Besides, except in Marxist mythology—which made "war the inevitable concomitant of capitalism"—nations seldom go to war with each other to increase the ratio of their export trade. The capitalists, thus maligned, had better sense than that. During the Algeciras crisis, for instance, the most determined advocates of a peaceable settlement were the representatives of the French and German coal industry, for they not only dreaded the price of war, they also were very much aware of the advantages which Franco-German cooperation would bring. When the final crisis came, as one historian of the period, the American scholar Bernadotte Schmitt, noted, "the loudest protests against war were made by the businessmen in Germany and Britain, who foresaw clearly what war would do to them."

How much more serious, in the actual event, were naval and colonial rivalries? True, the Germans might have done well to make Falstaff's motto their own. Discretion would have been the better part of valor. But not all was bluster. Germans and Englishmen, as noted, had agreed, by quiet negotiation, rather than by public collision, on the division of the Portuguese colonies. Some overall efforts to limit the naval race, such as that undertaken by Lord Haldane, then Secretary of War, during a mission to Berlin in 1912, ended in failure, but the two nations, by 1914, had managed to reach an informal understanding to construct new battleships at a ratio of Britain's 16 to Germany's 10. There was no reason why more instances of accommodation could not follow—not even the psychological barrier between the two nations. A look at some photographs of Edward VII enjoying a state visit plus family outing with his cousin William may suggest that more than party manners or the anticipation of a good meal lay behind the smiles and the bonhommie. True, such periods of contentment might alternate with others of utter distemper, but that was hardly a problem that applied to Anglo-German relations alone.

The love-hate relationship, for one, between the United States and Great Britain paralleled it rather closely. It offered the same spectacle of the transaquatic cousins who would now detest and now embrace each other, of the American who would now lionize his British visitor and now delight in twisting his tail, of the Britisher who would now envy those Americans for their wealth and now think them vulgar for it, of the American who would now make fun of the broad "a" and now join the English-Speaking Union. Yet, there was, in this period, neither war nor the threat of war between these two powers. But that was different, one might say. National interests coincided, so it was easy enough to keep whatever antipathies existed from influencing the conduct of foreign policy. Perhaps so, but to repeat the truism: nothing is forever in diplomatic history. There was no immutable rule which stated that Britain and Germany could never again inhabit a world in which their interests would once more run parallel. Psychological barriers, by themselves, certainly were not deep enough to prevent such a state of affairs, if only for the simple reason that if a good many Britishers still tended to dislike the Germans, there were others who detested the French or Russians even more.

What about the press and its role, then? The case here is strong. Who would not wish that editors and publishers had never discovered the charms that lay in fomenting trouble abroad. Yet there existed a mass audience avid for precisely this sort of entertainment. To have asked the owners of the press to refrain from supplying what their readers wanted—and what their own prejudices as often as not made them want as well—would have been expecting a great deal more wisdom and self-restraint than is given to the vast majority of us. Nor could the press, by itself, quite start a war. "Remember the Maine" was a one-time coup, never to be repeated. That the press lent a special element of danger to the ordinary conduct of diplomacy was true, yet when all the evidence is in, the influence of the press was substantially smaller than its own spokesmen thought. The final decisions continued to remain with the prime minister and not the night editor.

But were not chauvinistic editorials twice as perilous in a society that continued to accept war as a means of policy, that daily added to its arsenal of arms, and that had not enough statesmen of vision to keep the warmakers in check? Again, one wonders. For it was, and is, easy enough to point to the dangers that flow from the acceptance of war; it is rather more difficult to say what might be done to prevent the dangers that were, or are, likely to result from the abandonment of war as a means of policy. Would a nation announcing that it would under no circumstances have recourse to war truly be furthering the cause of peace—its own or that of others? Yet an argument against war as the ultimate resort of kings which fails to suggest what else the king might resort to will have to carry relatively little conviction. On the contrary, in the eyes of both statesmen and public, were not the preparations for war a true insurance against a war's actually taking place? Vickers and Krupp were not really kept as busy as they were turning out guns and shells because the men ordering these arms wished to use them against any live targets; rather each side wished to convince the other of their strength and seriousness of purpose, and hence of the unwiseness of deciding in favor of armed conflict. *Si vis pacem, para bellum*. "If you want peace, be ready for war," the Romans had thought. Deterrence, by many another name, is rather a venerable concept. So is that of leading from strength. To quote Raymond Sontag: "Armed forces were not intended primarily for use in war; they were to bring victory without war by forcing rival states to give way without an armed encounter."

But what if the available statesmanship did not match the weapons of destruction? After all, it takes keen eyesight and nimble feet to sense just where the brink ends and the precipice begins. How could matters have ended well at a time when Bethmann was sitting in Bismarck's chair, and Grey was holding the office once filled by Castlereagh? The answer—at times even rhetorical questions deserve an answer—was that matters very possibly could have ended well, for not only is genius customarily in short supply, but the statesmen of Europe were, with hardly an exception, if not inspired, at least competent and often quite shrewd men. So were many of the people serving under them, especially the diplomats in the field. As one looks at the dispatches of Paul or Jules Cambon, or of others for whose mention it has been hard to find room in a necessarily concise narrative—Marschall von Bieberstein on the German side, for instance, or Sir George Buchanan on the British—one is tempted to think of the era as at least the silver age of European diplomacy. And these were, again almost without exception, men genuinely committed to the cause of peace. Count Pourtalès, Germany's ambassador to St. Petersburg, openly wept when all had failed and, his hands shaking, he delivered his country's declaration of war to a Russian foreign minister who, in reply, embraced the representative of the power that Russia was now pledged to fight to the finish. Pourtalès's colleagues might do better at controlling their public displays of emotion, but they would feel as he did. The people, when the end came, were jubilant; the statesmen remained sober.

How deep, then, the trouble? Deep, but not fatal. War was possible in 1914—when is it not?—but it was not inevitable. "You may see" wrote the Under Secretary of the Foreign Office in London to the British Ambassador in Berlin on May 4, 1914, "there is little of interest taking place at this moment in Europe, and were it not for the troubles in Mexico we should be in comparative calm here." Not everyone was equally cheerful in their appraisal, of course, but an anatomy of many another year, whether 1874, 1884, 1894, 1904, or 1994, if undertaken as mercilessly as the inquest into 1914 necessarily is, may quite likely uncover troubles fully as profound as those of 1914.

We do not, for obvious reasons, undertake such inquests, but we may say that, all things considered, war was no more foreordained for 1914 than peace was for 1994, or 2004. Too much was right with the world of 1914: the nations, on the whole, had learned to live with the differences that divided them; several of the outstanding points at issue between rival powers were being settled or mitigated; and as the changing relationship between Britain and France, France and Italy, or Russia and Britain had shown, nations that were very dissimilar in temper, ambitions, and historical memory were fully as capable of accommodation as they were of conflict. History is forever in flux. In time, the particular difficulties that beset Europe *circa* 1914 might well have vanished, changed, or looked rather insignificant when placed next to other and newer problems.

1939 is the exception and not the rule. It is rare that a power is truly bent on war, as Hitler's Germany then was. In 1914, that sort of will to war was absent, although error, miscalculation, and sheer accident were very much present. But let us, because they were present, and because the end result was catastrophe, not become too absorbed by a sweeping search for underlying causes. Let us avoid the fallacy of saying that because the First World War's consequences were so momentous, its causes must of necessity have been equally significant. In 1731, a Captain Jenkins told London of how an overly jealous Spanish search party had boarded his ship, hung him from the foremast, and cut off part of his ear. As a result of this newsworthy but not world-shaking event, England plunged into a war with Spain—appropriately called the War of Captain Jenkins' Ear—that lasted for two years. Or there was Frederick the Great's frank admission that it had really been nothing more than his youth, and "the satisfaction of seeing my name in the newspapers" which had in truth caused him to attack Maria Theresa in 1740 and to begin the long War of the Austrian Succession. Of course, other and deeper causes contributed to the origins of both of these wars, but so did these immediate reasons. Of course, there were many explosives about in early twentieth-century Europe—there seldom are not—but it still took a fire to light them, and certain quite unpredictable winds to fan the flames. We need, therefore, if we are to understand the war's origins, to turn to its direct source, the assassination of Archduke Franz Ferdinand of Austria-Este, and to the events that followed it.

# Chapter 4

# Catastrophe: Sarajevo, the July Crisis, and the Outbreak of War

# The Shots of June

## *The Preparations*

More than one lifetime after the double murder in Sarajevo, a fair number of essential facts surrounding the crime's origins and authorship still remain in doubt. Some of the participants have told stories in which they have freely mingled truth with fiction. Others have kept entirely silent. Others yet have been silenced. The account that follows can therefore claim no more than that it is based on the testimony of the more credible among the surviving witnesses, and that it appears to be the most likely among several possible versions. But with these reservations in mind, as we try to reconstruct the story, it would seem that at some point between late 1913 and early 1914, Dragutin Dimitrijević, a lieutenant colonel in the Royal Serbian Army, decided that he wished Franz Ferdinand, heir apparent to the throne of Austria-Hungary, to die.

Dimitrijević was an interesting person. The official position he held was that of chief of intelligence of the Serbian army. Among the unofficial activities to which he devoted himself were the affairs of a secret society named Union or Death, an organization more popularly known, to the extent that it was known, as the Black Hand. The aim of the Black Hand, as of several other groups working in harmony or in competition with it, was the creation of a Greater Serbia, to include Bosnia-Herzegovina to the west, and Macedonia to the southeast. Toward this end, the Black Hand sponsored a substantial amount of propaganda and subversion in those regions. Also toward this end, Dimitrijević resolved on the death of Franz Ferdinand.

Archduke Franz Ferdinand of Austria-Este was heir to a throne whose occupant, the Emperor Franz Joseph, was eighty-four years old and frail. Franz Ferdinand's accession to power might spell the end of Dimitrijević's hopes; too much had been reported about the Archduke's political plans and ideas. The precise shape these plans would have taken must of necessity remain conjectural, but it would seem that in one form or another, Franz Ferdinand was ready to grant some notable concessions to the empire's Slavs. One scheme discussed would have provided for the establishment of Trialism, with the allotment of a role to the Slavs comparable to that played by the Germans and Magyars

*Colonel Dragutin Dimitrijević, called Apis, in one of his rare photographs.*

*The chosen victim: Archduke Franz Ferdinand of Austria-Este, shown here with his wife, Sophie, and their children, Maximilian, Ernst, and Sophie.*

in the monarchy. Another involved the possible introduction of a more confederate system along American or Swiss lines. Whatever their specific shape, the archduke's projected reforms had to be anathema to the Black Hand. The last thing that organization wished to see in Bosnia was happy Slavs. The foundation of its work was a discontented minority.

Dimitrijević, then, had a fairly compelling motive for murder. He also had the opportunity. Franz Ferdinand, he learned, would be coming to Bosnia in June 1914 to attend some army maneuvers, at whose conclusion he would pay a state visit to the Bosnian capital of Sarajevo. In addition to motive and opportunity Dimitrijević had the means: a group of three young nationalist Bosnians resident in Belgrade, Trifko Grabež, Nedjeljko Čabrinović, and Gavrilo Princip, who were willing to kill and, if caught, take the consequences. After first being trained by the Black Hand in the use of firearms and bombs, the three were equipped with the requisite weapons (as well as with vials of cyanide, so that they could commit suicide once they had accomplished their mission) and smuggled back across the border into Bosnia, from where they openly traveled to Sarajevo.

Here, on the morning of June 28, 1914, they took up the stations planned for them along the route that the Archduke and his wife Sophie were scheduled to take on their trip through the city. It was a day of particular significance and bitter memories to any Serb, since it marked the anniversary of the fourteenth-century battle of Kossovo, or Blackbirds' Field, against the Turks that had brought an end to Serbian independence and placed the nation under Turkish vassalage. It had not been excessive tact that had guided the Austrians to choose this day of all days for the Archduke's visit. On the other hand, both the assassination and its details had been planned some time before the precise date of the Sarajevo visit was known. It was indignation directed in advance, not any sudden visions of Kossovo producing homicidal impulses too strong to resist, that threatened the Archduke.

The three young men from Belgrade were augmented by a group of four local associates, although "augmented" may be a misleading term, since none of the four matched them in either resolution or courage. They had, one strongly suspects, been placed there mainly in order to give the crime the flavor of a more home-grown affair.

Not one of the seven, by the way, might have been able to take up his post that morning had a political decision made earlier in Belgrade gone just a bit differently. The facts—to the extent, again, that any "fact" about Sarajevo can truly be asserted—were these: some time in May, Nikola Pašić, Serbia's prime minister, learned about Dimitrijević's plot. This knowledge confronted him with a truly unenviable choice. If he were to tell the Austrians nothing and allow the assassination to proceed, he risked incurring the gravest of diplomatic crises and quite possibly war. If, on the other hand, he were to warn Vienna in an effort to prevent the plot from succeeding, he would in so doing have to admit just what sort of anti-Austrian activities had been tolerated on Serbian soil. In fact, he would also risk incurring the open enmity of the Black Hand at a time when his government's relations with that organization happened to be strained enough as it was. A possible third choice, the dismissal of Dimitrijević and the suppression of the Black Hand, apparently never struck Pašić as at all realistic, an estimate in which he very probably was right.

The compromise Pašić chose in the end, considering the dilemma that he faced, was as bad as it was comprehensible. He instructed the Serbian envoy in Vienna to look up the Austrian minister in charge of Bosnian affairs and suggest to him, in very general terms, that he consider a cancellation of the archducal visit. This, early in June, was what the envoy proceeded to do. Given the discontent among Bosnia's Serbs, he said, might it not be wiser to have the maneuvers take place in some other province of the Empire and to have Franz Ferdinand refrain from visiting Sarajevo. No, no, said his kindly and guileless Austrian host, there was no need to worry; according to his information, things had been very peaceful in Bosnia recently. This ended the conversation. The envoy's message had been so lacking in specific information, and the minister hence so unmoved by it (although he did try to humor his visitor by saying, as he showed him to the door, "Let us hope nothing does happen") that neither Franz Ferdinand nor the police were as much as informed of the envoy's call. Nor did Belgrade, even though aware of what had, or rather had not, taken place, ever revert to the matter.

That the Austrians, then, were not even conscious of the fact that they had been warned was perfectly understandable. What was less so was that the security precautions taken in Sarajevo during Franz Ferdinand's visit were at best routine, and at worst, careless. Of the thousands of troops in the area for maneuvers, none were used to guard the streets. Instead, all security arrangements were left in the hands of the local police, who numbered no more than a hundred and twenty in all. This included uniformed police and plainclothes detectives, and while perhaps the force was not inordinately incompetent, they were certainly no more efficient than an average provincial police force—whether in Sarajevo, Bosnia, or Dallas, Texas—is likely to be.

## *Sarajevo, June 28*

Even so, what followed that fateful Sunday, June 28, 1914, was not foreordained. The day's program for Franz Ferdinand and his party comprised first a reception at the city hall, followed by a visit to the local museum, then lunch at the governor's residence, and after that, the departure for Vienna. At about 10 A.M., the visitors set out on their way to the city hall. They were traveling in a number of open touring cars, with Franz Ferdinand, his wife Sophie, and the governor of Bosnia, General Potiorek, riding in the second car. They passed the first conspirator, one of the locally recruited boys, who in a flash decided that dearly as he wished to see Franz Ferdinand dead, he rather fancied the idea of staying alive himself; hence he did absolutely nothing to disturb that arrangement. A minute or so later, the cavalcade passed the second conspirator. He was Nedjeljko Čabrinović, one of the Belgrade-trained group. Čabrinović drew his bomb from his pocket, struck its percussion cap against a lantern post, took careful aim at Franz Ferdinand's general's helmet with its green panache, and let go.

Two things happened next. One was that the driver of the archduke's car, seeing the black object hurtling at him, instinctively pressed his foot down on the accelerator, so that, with the car speeding forward, the bomb did not directly strike its intended victim. The other was that Franz Ferdinand, having also caught sight of Čabrinović's missile, raised his hand to protect his wife. With this motion, he chanced to deflect the bomb, which fell into the street directly behind his car. There it exploded, injuring a number of spectators lightly and giving the governor's aide, riding in the next car, a head wound that looked serious enough to require his being taken to the hospital.

The plot, so painstakingly prepared by Dimitrijević and his friends, appeared to have failed. Within minutes, Čabrinović was under arrest. The cars went on their way again, and not one of the remaining five conspirators that they passed had either the courage or the presence of mind to raise a hand against Franz Ferdinand.

At the city hall, once the formal reception was over ("All the citizens of the capital city of Sarajevo," said the mayor solemnly as his archducal guest fumed, "find that their souls are filled with happiness, and they most enthusiastically greet your Highnesses' most illustrious visit with the most cordial of welcomes") Franz Ferdinand, his aides, and the representatives of the local authorities went into a quick conference to decide how best to alter the remaining program for the day. One narrow escape was quite enough; there was no need even for an officer and a Habsburg to take additional risks.

Hurriedly, they considered, and rejected, three suggestions that in the event would very likely have saved the archduke's life and prevented a world war. One was to cancel the museum visit and to drive directly to the governor's residence. Another was that to be even safer, they drive back, at high speed, to Franz Ferdinand's hotel outside Sarajevo. Franz Ferdinand rejected both. He did not mind making some concessions for the sake of security, but he felt that he must visit Potiorek's wounded aide at the hospital before leaving Sarajevo. The next suggestion someone then put forward was to wait at the city hall until two companies of troops could be brought into town from the maneuver area and the streets cleared of spectators. This one was rejected by the governor. The troops, said Potiorek, were not in proper uniform to be lining the streets.

Still, the arrangements finally agreed on were sensible enough. The group would avoid the narrow side streets of the original route and instead drive first to the hospital by way of Sarajevo's main avenue, the Appel Quay, and then continue on to the museum by an equally unscheduled route. The plan might well have worked had it not been for two factors. One was that at the time the discussion was going on at the city hall, Princip crossed from the sun-drenched river side of the Appel Quay, where he had been standing, to the shaded side of the avenue and strolled a few steps down Franz Joseph Street, a shopping street that led off the Quay. The second was that no one informed the driver of Franz Ferdinand's car about the changed plans.

What happened next was this: Instead of driving straight down the Appel Quay, the car carrying Franz Ferdinand, his wife, and the governor turned into Franz Joseph Street, precisely as the chauffeur had originally been instructed to do. As it did so, Potiorek, realizing the error, leaned forward and called, "What is this? This is the wrong way! We're supposed to take the Appel Quay!"

The driver (a general's orders are a general's orders) put his foot on the brake and began to back up. The maneuver put the car in a straight line with Princip. Princip stepped forward, drew his gun, and from a distance of not more than five feet, fired twice. No amount of planning could have provided him with a better target. One bullet pierced Franz Ferdinand's neck, while the other entered Sophie's abdomen.

The Duchess, not wholly aware perhaps in the first second or so of what had been taking place, but seeing blood spurt from her husband's mouth, called, "For heaven's sake! What happened to you?" and slipped from her seat, her face falling between the Archduke's knees. Franz Ferdinand, turning to his wife despite the bullet in his neck, said "Sophie dear! Sophie dear! Don't die! Stay alive for our children!" Fifteen minutes later, both were dead.

## THE ECHO IN VIENNA

Within hours, extra editions of the Viennese papers carried the story from Sarajevo. The news caused more political anger than personal sorrow. Franz Ferdinand had not been a widely beloved man—"nothing in his life became him like the leaving it"—but the political implications were obviously ominous. Initial information about the assassins might be scarce, but what was known was that they were ethnic Serbs, and they had committed their act in politically restive Bosnia with its border toward Serbia, facts quite sufficient to throw strong suspicions on Belgrade. What was to be done now?

The men in charge of conducting the policies of Austria-Hungary were not sure. They were not exactly unaccustomed to Serbian-supported acts of terror, propaganda, and sabotage on Austrian soil, but this brazen act was something new, provided that the Serbian government could in fact demonstrably be connected with the two murders. Could it? Was the crime different in nature from previous provocations? Should Austria-Hungary now resort to arms, as some among the military had repeatedly suggested ever since the Annexation Crisis, even though Franz Ferdinand, on that and subsequent occasions, had consistently argued against war? The government did not quite know. Hence they would explore two matters first. One was Germany's attitude. Would Vienna be able to count on Berlin in case an Austro-Serbian conflict widened into general war, as well it might, since Russia could conceivably come to Serbia's aid? The other was the question of evidence. To pursue it, a senior official of the foreign ministry, Friedrich von Wiesner, was charged with an investigation of the assassination. Would he ascertain what proof, not suspicion but proof, there was to connect the assassins with Belgrade?

By themselves, these reasons for delay were far from bad. Austria, unaided, was in no position to win a war against a Russo-Serbian coalition. Even against Serbia alone, it now turned out, the army was unprepared to strike quickly; mobilization would require at least two weeks. Also, it would be of considerable advantage to adduce proof of Belgrade's culpability before making the first overt move. (The United States, forty-eight years later, waited very patiently until it could substantiate its charges against the Soviet Union before imposing its blockade on Cuba.) Yet in a way, a discussion of the merits of Vienna's motives for procrastination is irrelevant, for the delay, no matter how defensible its reasons, turned out to be a disaster. In retrospect, it is obvious that if Austria had acted forcefully and decisively against Serbia on June 29, when the impact of the double murder was at its sharpest, the sympathies of the world would have been with the bereaved Dual Monarchy, and the very worst that might have happened would have been a limited war between Austria-Hungary and Serbia. Nor was it quite that hard to see at the time that this should be so. Had the

Austrian foreign ministry been in better hands, this very likely would have been the course followed.

## Germany's Blank Check (July 5–6)

A week after the murder, the Austrians had the answer to their first question. On Sunday, July 5, a special emissary sent by the Austro-Hungarian foreign minister, Count Leopold von Berchtold, arrived in Berlin, bearing a personal letter from Franz Joseph to William II. The Dual Monarchy, read the letter, (which was obviously inspired by Berchtold rather than by the privately much more reluctant Emperor) could withstand "the Pan-Slav flood" only if "Serbia's role as a power factor in the Balkans is ended." One would have to aim at the "isolation and diminution of Serbia." What would Germany's position be if Austria now followed such a course?

That morning the Austrian envoy presented the letter to William at the New Palace in Potsdam. The Kaiser at first avoided comment. Then, after a sociable lunch with his visitor, he abandoned his reserve. Would the envoy tell his sovereign that even in the event of "some serious European complications," Austria-Hungary could rely on "Germany's full support." If Russia, to put the matter more specifically, should come to the aid of Serbia, then Germany would aid Austria. In fact, asked William II, why not strike against Belgrade right now? As matters stood, the Russians were "hardly ready for war and would certainly think very carefully before resorting to arms."

In the afternoon, the German Chancellor, Theobald von Bethmann Hollweg, who had joined the others at the New Palace, confirmed his monarch's promise, and the special emissary was able to inform Berchtold of Germany's "blank check." The phrase was all too apt. The Austrians were free to fill in the amount on a form that the Germans had cosigned in advance.

There are several quite plausible explanations for this piece of German folly. The Emperor apparently thought that the Austrians would now proceed to act without further delay, and that, given the merits of their case against Serbia and Russia's relative unreadiness for war, whatever fighting there might be would be confined to Austria and Serbia. The German foreign minister, a less sanguine man, was not available for comment and counsel that weekend. He had recently gotten married and was on his honeymoon in Switzerland. The Chancellor, who was there, was full of forebodings but felt that Germany had no alternative but to say yes to the Austrians. "It is our old dilemma in any Austrian Balkan action," he grumbled to an aide a day later. "If we encourage them, they say that we pushed them into it. If we dissuade them, they say that we abandoned them." And everything considered, dissuasion was the worse evil, for "in that case, they will draw close to the Western powers, whose arms are open, and we will lose our last powerful ally."

These reasons deserve to be mentioned; it was neither utter blindness nor invincible malevolence that made the Germans act as they did. Yet in a sense, they are as irrelevant as are the explanations for Austria's procrastination. All these many years later, we are still staggered by the ineptitude of the German response. For in the first place, no one in Berlin seemed to have asked the question of what really might be an acceptable price for maintaining the alliance with Austria-Hungary, an empire plagued from within by the apparently perpetual conflicts among its various nationalities. And besides, even if that empire's health had been entirely sound, the risk involved in saving the alliance—the risk, not the intent—was nothing less than the destruction of the European peace, and with it, of Imperial Germany itself. That risk was out of all proportion to the possible gain, and it did not take hindsight to know this. Bethmann's aide, after recording the words just cited, went on to note in his diary: "An action against Serbia could lead to a world war. The chancellor expects that such a war, no matter how it might end, will revolutionize all that exists."

Yet Bethmann was willing to take the chance, was ready to play with fire for dubious gain. For the aide who recorded the Chancellor as saying, at the same time, that "should war come from the East, . . . we have a chance of winning it. Should war not come, and a worried Tsar connect peace to France, that will give us an opportunity to break up [maneuver apart was the precise phrase he used] the *Entente*."

This kind of risk-taking befitted a gambler more than a Chancellor. Twenty-five years earlier, Bismarck had described what was at stake in unmistakable terms. "Only some complications between Austria and Russia could involve Germany in a war with Russia," he had written to his king. "Since the cost of such a war, even under the most favorable of circumstances, would be unacceptable for Germany, we must endeavor to do what we can to prevent the Austrian war."

Bethmann had done far less than that. The ultimate decision now lay with Austria.

## "There is Nothing to Indicate . . .": The Wiesner Investigation (July 13)

At the foreign ministry in Vienna, Berchtold was delighted with the German promise. It was now or never, he thought, that the Habsburg monarchy should move against Serbia. Twenty-four hours later, however, there arose a formidable obstacle to his plans. In a council of ministers held on July 7, Count Stephan Tisza, prime minister of the Hungarian half of the monarchy, sharply opposed war or any course of diplomatic action that might lead to war with Serbia.

Tisza was a devout Calvinist who abhorred the use of force. "War," he wrote in a family letter in August 1914, "even if victorious, is terrible. To my soul, every war means misery, anguish, devastation, the shedding of innocent blood. . . ." Tisza was also a working Magyar politician who had no interest at all in "diminishing" Serbia. Hungary had enough dissatisfied minorities within its borders; he wished to acquire no more. It would be a grave mistake, he told the council of ministers, to force the issue with Serbia. A European war would be a "terrible calamity." And by the way, he would like to inquire where the Sarajevo police had been that bloody Sunday. It surely was "an unspeakable state of affairs if six or seven characters armed with bombs and guns could line the route of the murdered Heir to the Throne and not have the police observe or remove a single one of them."

Berchtold and his friends labored hard to change Tisza's attitude. An investigation of the Sarajevo police would have to wait, they said. What mattered now was to strike against Serbia; any "policy of hesitation or weakness" might lose Austria-Hungary the future support of Germany. (The irony was unconscious, but both the Germans and the Austrians were talking themselves into risking their own destruction for fear of losing one another as an ally.) In the end, Tisza gave in, though only on condition that the monarchy would not annex "an inch of Serbian territory." The concession was readily granted, since it was meant to be broken from the moment it was made.

It had taken a week to persuade Tisza. Meanwhile, more precious time had been lost. World opinion was recovering from the initial shock of the double murder.

Perhaps it might be possible to revive some of the lost indignation by providing new details about the crime and, above all, by uncovering some clear evidence that would connect official Serbia with the plot.

The person entrusted with that task had been Friedrich von Wiesner of the Austrian foreign ministry. To follow his investigation, we need to retrace our steps to June 28 for a moment.

Immediately after Čabrinović's attempted, and Princip's successful attack, the two young men followed the instructions they had received in Belgrade and attempted to commit suicide. It was Colonel Dimitrijević's only really reliable assurance that, if arrested, they would not talk. However, no murder planned by remote control is ever likely to be entirely perfect, and the cyanide capsules that Čabrinović and Princip managed to swallow as they were being arrested and led to the police station caused them some modest pain and discomfort, but did not prove fatal. For reasons unknown then, and likely to remain unknown forever, the poison they had been given was defective.

The Sarajevo police, then, had Princip and Čabrinović safely in custody; and in the days that followed, nearly all of the local fellow conspirators were arrested—some because they were routine suspects, others because one of Princip's friends talked too much. The authorities might now have found the evidence linking the assassins to Colonel Dimitrijević, had it not been for the extraordinary skill with which Princip and his associates fended off the questions they were asked about Belgrade, a skill impressive even in the face of a certain lack of imagination with which the investigating judge was handling their interrogation.

Even so, the young men could not hide the fact that Princip and Čabrinović had been recent visitors to Belgrade, nor could they very well deny that the bombs used had been of Serbian manufacture. (The guns were Belgian army issue.) In addition, the names of two men who had aided the assassins in Belgrade finally emerged from the questioning. They were those of Milan Ciganović, a Serbian railroad employee, and of Voja Tankosić, a major in the Serbian army. But what the Sarajevo authorities did not discover in this connection was as important as what they did. They never learned that Tankosić was Colonel Dimitrijević's Black Hand aide; in fact, they were to remain wholly unaware of the existence of the colonel and his organization.

On July 10, von Wiesner arrived in Bosnia. He had studied all the material available to him in Vienna and found it of little profit. He now spent day and night in Sarajevo conferring with the authorities and going over the files they had been accumulating. On the morning of July 13, after four hours of sleep, Wiesner drafted the dispatch for which Vienna had been waiting.

The body of his report mentioned the links he had been able to establish between the crime and Belgrade: the bombs, the assassins' associates Ciganović and Tankosić, and the illegal border crossing that might not have been feasible without the aid of several Serbian frontier officials. The really crucial part of Wiesner's report, however, consisted of two short and fateful sentences, which he was to rue for the rest of his life. "There is nothing to indicate, or even to give rise to the suspicion," he wired, "that the Serbian government knew about the plot, its preparation, or the procurement of arms. On the contrary, there are indications that this is impossible."

This, then, was the telegram for which the Austrians had been willing to postpone immediate, drastic action. Now that it had arrived, Vienna was worse off than it had been on June 28. Berchtold and the war party still felt that no matter what Wiesner might say, the time had come to move against Serbia—but where was the evidence that would convince anyone but the converted?

It was the triumph of Dimitrijević's planning. Nor would the Austrians learn much more when the trial of the assassins was held in October. The defendants, on that occasion, handled themselves with the same courage, intelligence, and determination they had shown all along.

When sentence was pronounced, neither Princip nor Čabrinović received the death penalty. Both were under twenty, and thus under Austrian law exempt from that punishment. Having passed that crucial birthday, some of their helpers were less lucky, and Princip and Čabrinović themselves were to die in prison of tuberculosis a cruelly short time before the end of the war. Few members of the conspiracy survived them by much. Dimitrijević, in fact, preceded them in death. In 1917, he was tried and executed by his own government, ostensibly for engaging in one plot too many. Only two of the assassins managed to survive for long. Both were among the local boys who had joined the conspirators from Serbia on June 28. In fact, both went on to long and distinguished careers, one as a noted historian at the University of Belgrade, the other as the curator of the Sarajevo museum he had helped to prevent Franz Ferdinand from visiting that Sunday.

## No More Cause for War? Austrian Ultimatum and Serbian Reply (July 23–25)

Evidence or not, Berchtold, in mid-July, was resolved on humiliating Serbia. An unrestrained Serbia, he felt, posed a vital threat to the

Habsburg monarchy, no matter who had sent Princip to Sarajevo. If there was no proof linking Pašić's government to the assassination, that was a pity, but Sarajevo, after all, was merely one provocation among many. He would, in one last gamble to save the monarchy, confront Serbia with a number of all but unacceptable demands; and when these had predictably been refused, he would set the army in motion against Belgrade.

What would Serbia's allies do? The answer arrived even before Berchtold's ministry was finished drafting its ultimatum to Serbia. Supposedly, the ultimatum was being prepared in the strictest secrecy, and the Austrians certainly took great pains not to frighten their German ally with any advance information about its contents. But rumors about what was going on in Vienna were spreading. On July 18, Sergei Sazonov, Russia's foreign minister, thought it necessary to take the Austrian ambassador aside and tell him that "Russia would not be indifferent to any effort to humiliate Serbia. Russia would not permit Austria to use menacing language or military measures against Serbia. In short: '*La politique de la Russie est pacifique mais pas passive.*'"

But did "pacific but not passive" really mean that Russia would dare to move? Austria had Germany's blank check; how could Russia cope with an Austro-German coalition? The answer came in terms as clear as those of Sazonov's warning. On July 20, Raymond Poincaré, President of the French Republic, arrived in St. Petersburg for a state visit. The next day, during a reception of the diplomatic corps, he asked the Austrian ambassador if he had any news of the Serbian affair, and on receiving an evasive answer, said: "With a little good will, this Serbian business is easy to settle. But it can just as easily become acute. Serbia has some very warm friends in the Russian people. And Russia has an ally, France."

Work on the Austrian ultimatum continued as though neither Sazonov nor Poincaré had spoken. On July 23, it was finished. At 6 P.M. it was presented in Belgrade. It was a brutal document whose directness had few precedents in diplomatic history.

The Belgrade government, read the ultimatum, was to put an immediate stop to all Serbian sponsored subversive activities on the territory of the Austro-Hungarian monarchy, as well as to all anti-Austrian propaganda in Serbia itself. As a specific instance, the *Narodna Odbrana*, or *National Defense*, a secret organization pursuing Greater Serbian aims, was to be disbanded. Judicial proceedings, the ultimatum went on, were to be instituted against any person in Serbia who had been an accessory to the Sarajevo crime, with "agencies delegated by the Royal and Imperial

[Austrian] Government" taking part in any investigation that might, in this connection, be undertaken in Serbia. Tankosić and Ciganović (the two Serbian helpers mentioned by the assassins) were to be arrested "forthwith." The Serbian government was to cooperate in putting an end to the illegal traffic in arms across the border, and the Serbian officials who had aided Princip in crossing the border were to be punished.

The Serbian government had precisely forty-eight hours in which to decide whether to accept or reject these demands: "The Austria-Hungarian Government expects a reply of the Royal Government by 6 P.M., Saturday, July 25 at the latest."

Good as Austria-Hungary's basic case against Serbia was, it had been all but destroyed by the ultimatum's text and timing. For all the ultimatum's harsh language, there was no solid evidence linking the Serbian government to the assassination. Tankosić and Ciganović were minor figures. Mention of the *Narodna Odbrana* was next to meaningless, since that organization, while once indeed as menacing as the Black Hand, had meanwhile turned respectable and by 1914 bore roughly the same relation to Colonel Dimitrijević's group that Kropotkin's militant anarchism bore to Bakunin's peaceable one. Fine points, perhaps, that might have gone unnoticed, had not so much time been allowed to pass since June 28. What people saw at the end of July 1914 was not an archduke gasping "Stay alive for our children's sake" to a wife who was bleeding to death as fast as he was, but a large nation bullying a small one.

It was hardly strange, then, that when the powers learned of the ultimatum to Serbia on the morning of July 24, they did not rush to Austria's side. The Russians, in fact, quickly did the opposite. In a meeting held that afternoon, the council of ministers gave its approval—which the Tsar endorsed a day later—to a proposal submitted by Sazonov stating that if such a move were to appear useful, several Russian army corps be mobilized against Austria. The British were more pacific. Sir Edward Grey proposed a conference between the four major nations not directly involved—Germany, Italy, France, and Great Britain—to mediate the conflict. Other capitals withheld action for the time being. All now realized that Europe was facing a major crisis, but before making any irrevocable decisions, they would await the Serbian reply.

On July 25, at 5:58 P.M.—two minutes before Vienna's ultimatum was about to expire—the Austrian envoy in Belgrade received that answer. In tone and content, it was remarkably conciliatory.

The Serbian government, the note promised, would, as soon as parliament met, introduce a press law providing the most severe penalties for any "incitement to hatred and contempt of the Austro-Hungarian Monarchy." It would dissolve the *Narodna Odbrana* even though, as the Serbians dryly noted, the Austrians had provided them with no proof of any illegal action on the part of that organization. It would "severely punish" the frontier officials who had allowed the Sarajevo assassins to cross into Austria. It would "reinforce and extend the measures which have been taken for preventing the illicit traffic in arms and explosives across the frontier."

In fact, the Serbian note agreed to just about all of the demands of the Austrian ultimatum. There were only two exceptions. One concerned the arrest of Ciganović, the other the investigation into the origins of the assassination. Tankosić, said the note, had been taken into custody, but Ciganović, unfortunately, had managed to evade arrest. As for any participation of Austro-Hungarian agencies in investigating the crime on Serbian soil, the government regretted that it was unable to "accede to such an agreement, as it would be a violation of the Constitution and of criminal procedure."

Compliance on either point might, of course, have meant disaster to the Serbians. Belgrade was neither in a position to permit an investigation that might uncover a trail leading to the chief of Serbian army intelligence, nor could it afford to have the Austrian police interrogate Ciganović. Ciganović, who held an Austrian passport and whose extradition, unlike Tankosić's, could therefore not very well be denied, had accordingly been warned some weeks before to vanish from sight; and while the precise circumstances of his disappearance may never become public knowledge, there is some room for impugning the zeal with which the Belgrade police set about the task of finding him.

At the time, both seemed minor points. The mood, rather, was one of relief at Serbia's concessions, apparent and real. William II expressed the feelings of many when he wrote after reading the note: "A brilliant performance for a time limit of only 48 hours! A great moral success for Vienna, but with it, all reason for war is gone. . . ."

He had misread Austria's intentions even more badly than Serbia's reply. For the Austrians fully intended to take offense, no matter what Pašić might say to them. Within minutes of being handed the Serbian note, and after no more than a glance at its contents, the Austrian envoy told Pašić that he considered the document unsatisfactory and that he, therefore, had "the honor to inform Your Excellency . . . that

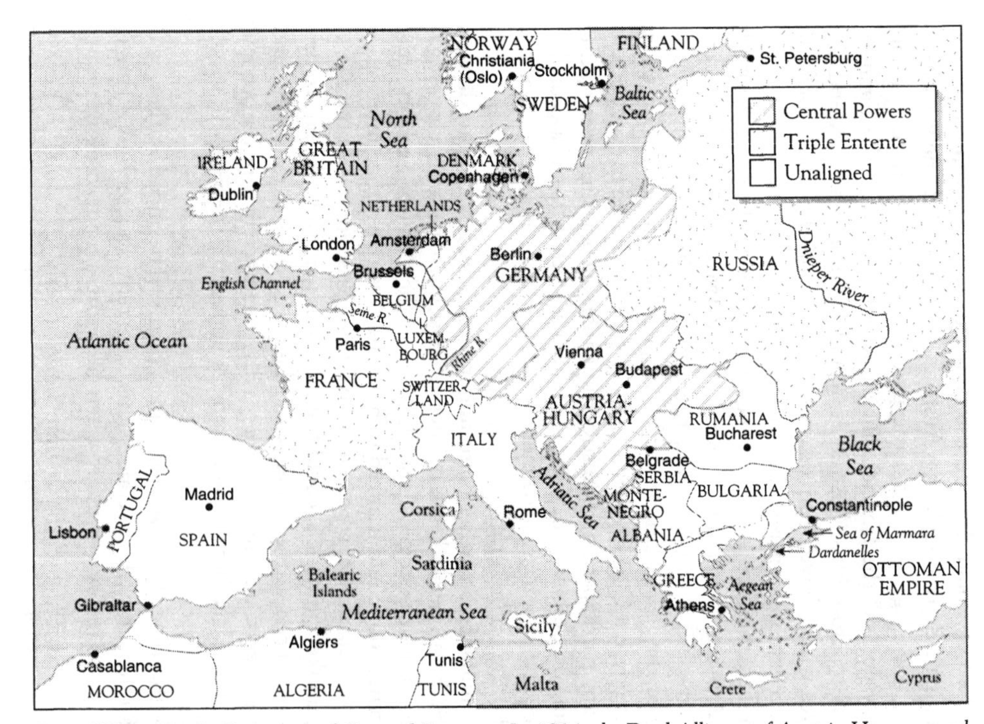


Eve of War: Triple Entente and Central Powers. *In 1914, the Dual Alliance of Austria-Hungary and Germany—the so-called Central Powers—faced the Triple Alliance of France, Russia, and Great Britain. (Italy, nominally still allied to the former, would first stay neutral, then join the Entente in 1915.)*

from the moment this letter reaches Your Excellency the rupture in the diplomatic relations between Serbia and Austria-Hungary will have the nature of a *fait accompli.*" At 6:30, 32 minutes after the receipt of the note, the envoy and his staff were at the Belgrade terminal, catching the train home. Three-and-a-half hours later, the Austrian chief of staff received Imperial orders to mobilize the armed forces of Austria-Hungary against Serbia.

## PEACE OR WAR? THE SMALL WAR BEGINS (JULY 25–28)

Europe began to prepare for the worst, while still hoping for the best. Earlier on July 25, the Tsar had given his approval to Russia's partial mobilization. (The extent to which this decision strengthened Pašić in his resolve to resist full compliance with the Austrian demands has remained subject to much scholarly debate.) Later that day, the French, very discreetly, lest they be accused of aggressive intentions, took some precautionary military measures. A day later, the admiralty in London ordered the British fleet, which had been assembled in review and was ready to disperse, to remain at its stations. But at the same time, the statesmen were trying for peace, or at least for a localization of the conflict.

The Germans concentrated on ways of restricting the conflict, if war should come, to the original parties to the dispute, Austria and Serbia. The Tsar had the sudden inspiration of submitting the whole matter to the International Court of Justice at the Hague. The more practical British—the Tsar's suggestion, in fact, never passed beyond Sazonov's desk—reverted, in a more precise form now, to their earlier idea of a four power conference: let the ambassadors of France, Germany, and Italy in London, said Sir Edward Grey on July 26, meet with him and search for ways to resolve the conflict. The Italians promptly accepted the proposal. So did the French, although with a notable lack of enthusiasm. The Germans alone, on July 27, refused. It was their greatest blunder since the day of the blank check; they were still under the illusion that they could localize the conflict.

It did not take them long to recognize their error. Later that same day, after a hard look at the diplomatic dispatches that had been pouring in, the Germans faced the bitter truth that if Austria and Serbia were to fight, so would Russia, France, and quite possibly Great Britain. Now the Germans radically altered their course. Might it not be wise, they urgently asked Vienna on the evening of July 27, to accept the Serbian reply as a basis for negotiation?

It was too late. Come what may, the Austrian government was resolved to end what it considered the Serbian threat to the empire's survival. On Tuesday, July 28, without any further consultation with Berlin, the Habsburg monarchy declared war on the kingdom of Serbia.

## "Mobilization Means War": The Intervention of Russia (July 25–29)

The small war had begun. Frantically, Europe's leaders tried to prevent a larger one.

The historian, in recounting their efforts, is aware of a basic problem of historiography, which is that any comprehensible narrative of what took place in the ministries and palaces of Europe during the days following July 28 must of necessity contain some element of falsehood. For what in print is likely to look neat and logical, in truth, more often than not, was chaotic or at best improvised. Decisions were made with no time for proper reflection, messages crossed each other, and some of the most fateful errors were committed from motives no more profound or sinister than lack of information or sleep. It is with this in mind that we should view the events that occurred between the last days of July and the opening days of August 1914.

One of the most sensible proposals for saving the general peace now came from the Kaiser. In normal times, William II could be insufferable, a crowned buffoon without peer; in this moment of crisis, he was acting with as clear a head as anyone. Let the Austrians, he proposed on July 28, halt their armies when they reached Belgrade, and suspend the war. With Serbia's capital as a "pledge," the Austrians could be sure of obtaining some genuine satisfaction from Pašić's government. Meanwhile, there would be a breathing spell, and the powers would be able to gather for sensible peace talks.

Grey rather found himself in sympathy with the suggestion, but Bethmann failed to urge it on the Austrians with anything like real vigor. Berchtold remained silent, nor did he respond when Bethmann, on July 29, grew more insistent. The time was past when anyone could tell the Austrians just what they might or might not do about the Serbian issue, even if, as now seemed more apparent by the hour, this meant incurring the gravest risk of Russian intervention.

To prevent this sort of extension of the war, the Kaiser, on July 29, sent a telegram to his cousin, the Tsar. Would "Nicky," said "Willy" (salutation and signature are those of the text) realize that he, his Berlin friend and relative, was exerting his "utmost influence to induce the Austrians to arrive at a satisfactory solution with you. I confidently hope that you will help me in my efforts." Nicholas II, it so happened, had been sending a very similar appeal to the Kaiser at just about the same moment, begging him, "in the name of our old friendship," to do what he could "to stop your allies from going too far and to avoid a calamity such as a European war." William II quickly replied that he would, but that one obstacle lay in Russia's apparent military preparations. Berlin was hearing rumors about a Russian mobilization against Austria; if true, the Kaiser warned, this might end all hopes for peace.

It was a point on which the Tsar was willing to accommodate his cousin. But in so indicating, he made a fatal admission. "The military measures which have now come into force," he wired back during the night of July 29 to 30, were no sudden anti-Austrian gesture, but had been "decided on five days ago for reasons of defense on account of Austria's preparations."

The Kaiser was furious. "My work," he said, "is at an end." Here he had been trying to mediate, and what had the Russians done but use his efforts as a cover for obtaining a five-day start in military preparations. Worse was to come. To understand it, it is necessary to go back a few days and deal with the matter of Russia's mobilization in a certain amount of detail.

On July 24, a prominent German businessman and friend of the Kaiser, Albert Ballin, in London for a brief visit, was having dinner with two British acquaintances of his. They were the First Lord of the Admiralty, Winston Churchill, and Churchill's colleague in the war department, Lord Haldane. "I remember," Ballin said to them, "old Bismarck telling the year before he died that one day the great European war would come out of some damned foolish thing in the Balkans." The forecast seemed about to come true. Everything, said Ballin, depended on the Tsar now: What would Nicholas do if Austria were to chastise Serbia? "If Russia marches against Austria, we must march; and if we march, France must march, and what would England do?"

What the Tsar would do was suggest to the Serbians, in the aftermath of Berchtold's ultimatum, that while they should give no gratuitous offense to the Austrians, they might also rest assured that Russia would not remain indifferent to their fate. The Russians calculated that they could not afford a repetition of the Annexation Crisis, which had left the Serbians with the feeling of having been deserted by their Slavic protector. It was—or so it seemed to the men who made policy in St. Petersburg—a choice between taking the risk of war or accepting a major blow to Imperial Russia's prestige and power in the Balkans.

What the Tsar also would do, as we have seen, was to give his advance approval, on July 25, to Russia's partial mobilization against Austria in case of need. To this he added, on that same day, his agreement to issuing the necessary secret orders for a "period preparatory to war." Leaves were to be cancelled, reservists called, Russia's harbors closed—in short, the country would begin to prepare for mobilization. These were some of the measures that had so upset the Kaiser, yet they had seemed essential to Russia's military. War was a possibility, and Russia a vast country. Moving the Russian armed forces to their battle stations would take a very long time, and the nature of modern war was such that the power capable of striking the first blow would have an intolerable advantage. To recall Bret Harte's advice:

> Blest is the man whose cause is just;
> Thrice blest is he who gets his blow in fust.

Under normal circumstances, the Germans and the Austrians, with their better railroad system and their generally more efficient military machine, would have just that advantage. Hence, the need on Russia's part to begin preparing before they did.

On July 28, Sazonov decided that the time had come to go beyond the steps of the "period preparatory to war," and to cash in the Tsar's three-day-old blank check. Russia, he informed the powers that day, was beginning the partial mobilization of her armed forces toward Austria. Several considerations accounted for this decision. One was the Austrian declaration of war against Serbia, which had come after several urgent appeals of his to Berchtold for the opening of Austro-Russian talks had simply gone unanswered. Another was a French reassurance of support. In this moment of supreme danger, the French were doing no more to restrain their Russian ally than the Germans had done to hold back the Austrians.

There were, in the French case, more extenuating circumstances than in that of the Germans. Poincaré, President of the Republic, and Viviani, prime minister and foreign minister as well, were at sea, returning from their Russian state visit, and out of regular communication with Paris and Petersburg. "Not to have exact information about anything," Poincaré later wrote about the voyage,

> not even to have the essential data in hand as to the problem that was to be resolved, caused M. Viviani not only mental but physical suffering. He paced the *France's* deck in his agitation, silent for long periods, then returning at intervals affectionately to pour his anguish into my ear.

In Paris, Bienvenu-Martin, minister of justice and acting foreign minister in Viviani's absence, knew little of what was going on, and did not really understand the true seriousness of the crisis. In Petersburg, Paléologue, the French ambassador, and a strong advocate of the Franco-Russian alliance, was pursuing a policy of his own. He deliberately misinformed Paris about what was taking place in Russia. He also gave promises of support to the Russians which exceeded his instructions. This fairly complex background, however, did not particularly concern or interest Sazonov. What did concern and interest him was what France's official spokesman at Petersburg told him, and that was clear enough. To quote from the diary entry for July 28 of Baron Schilling, Sazonov's *chef de cabinet:* "On the instructions of his Government, the French Ambassador acquainted the Foreign Minister with the complete readiness of France to fulfill her obligations as an ally in case of necessity."

That leaves one more reason which deserves to be mentioned to account for Russia's decision to begin mobilization: Sazonov, while no fool, was a man given to moods, and worse, to gambles and to dubious judgments, and he may well have thought that partial mobilization would prevent general war, since it would frighten the Austrians without threatening the Germans.

Things would not work that way. Partial mobilization, he was to learn apparently within hours of deciding on it, was an illusion. The sad and simple fact was that the Russian general staff had no plan for it.

The partial mobilization that Sazonov had in mind would have placed four Russian military districts facing Austria on a war footing. This would have been a provocation. (The Austrians, after all, had restricted their military moves to Serbia and been careful to take no preparatory measures against Russia.) But given its limited scope, it might have been a tolerable provocation. "Might have been," for what now came to light was that the Russian general staff had failed to make any provisions for a limited mobilization against Austria alone. It had not been able to conceive of a war that would not be fought against both of the Central Powers; hence, all its plans provided for general mobilization, that is for mobilization against Germany as well as against Austria.

Then why had the army not spoken up, and made that point in the council of ministers of July 24 where partial mobilization was first tentatively decided on? It had not, because the chief of the general staff, General Janushkevich, was relatively new to his post—he had assumed it only five months previously—and was still unfamiliar with his office's mobilization plans. It was only after that meeting that some of his subordinates, better informed than he, told him that partial mobilization was next to impossible. It would have to be improvised (a difficult thing to do with a railroad timetable) and lead to nothing but confusion and disorder. It would have to be general mobilization or nothing.

When Sazonov now presented his decision for partial mobilization, Janushkevich had to tell him the bitter truth. No matter what the foreign minister may have thought privately about the caliber of his generals, it seemed too late for him to draw back. Had he known on July 24 that partial mobilization was a myth, he might have chosen a very different course of diplomatic action. But he had not known, and by July 28, he felt irretrievably committed to a policy of strength. He therefore did nothing to counteract the pressure that the military proceeded to put on the Tsar to agree to general mobilization.

On July 29, they had the agreement they wanted. That morning, the Tsar signed the order for general mobilization. Fortunately, it did not go out immediately. Under the rules, the ministers of war, the navy, and the interior were required to countersign it. Prenuclear societies, too, took their precautions, although they could be more leisurely about it, and it was not until evening that the three men had been located. All signed reluctantly. The minister of war knew full well how sadly prepared his army was. His colleague in the navy department realized that his fleet was no match for Germany's. The minister of the interior sensed that war for Russia might be the prelude to revolution. But, making the sign of the cross, he, as the others had done, signed.

What they also knew was that they had, in effect, been putting their signatures to a declaration of war. No international law specifically stated that mobilization automatically equaled war. It did not have to; the implications were clear.

In 1892, after signing the Military Convention that had just been concluded between his country and Russia, the French delegate, General Boisdeffre (he was, at that time, the deputy chief of the French general staff) had had this to say to the Tsar: "Mobilization is a declaration of war. To mobilize is to oblige one's neighbor to do the same. Mobilization causes the carrying out of strategic transport and concentration. Otherwise, to allow a million men to mobilize on one's frontiers without at once doing the same oneself is to forfeit all possibility of following suit, is to put oneself in the position of an individual with a pistol in his pocket who allows his neighbor to point a weapon at his head without reaching for his own." To which Nicholas' predecessor, Alexander III, had replied: "That is how I too understand it."

Yet there was still time to put down the weapon. At this point, just about everyone, it seemed, was warning everyone else in Europe to do so. Lord Grey, that Wednesday, July 29, was telling the German ambassador in London that if Germany and France were to be drawn into the Austro-Serbian conflict, England might be unable to stand aside. The Russians, with their military measures, were putting the Austrians on notice. The Germans, by means of the Kaiser's latest "Dear Nicky" telegram, were telling the Russians that their military preparations were jeopardizing the peace. What was more, one person was actually listening. On the evening of July 29, a moment before the orders he had signed that morning were about to go out, the Tsar instructed his generals to cancel general mobilization, and to replace it—whether this be practicable or not—with partial mobilization again.

## Finding, Suicide; Motive, Fear of Death— A Time of Ultimatums (July 30–31)

That was on Wednesday. On Thursday, July 30, the Tsar had changed his mind once more. Both his generals and his foreign minister had given him no peace. What the hour demanded, they insisted, was general mobilization, not partial measures. "Think of the responsibility you are asking me to take!" said Nicholas II. "Think of the thousands and thousands of men who will be sent to their death!" He was thinking of them, said Sazonov, but diplomacy's work was done; war was coming. What mattered now was that Russia be prepared.

Late that afternoon, the Tsar surrendered. Sazonov rushed to another room to telephone the chief of staff with the news. "Now," he ended their conversation, "you can smash your telephone. Give your orders, General—and then disappear for the rest of the day."

General mobilization was on again, and no telephone, smashed or whole, would ring to announce any orders to undo it. Most writers have been free with their criticism of Nicholas II for his behavior in this time of crisis. But only an extraordinarily strong man could ultimately have resisted the massed opinions of both diplomatic and military experts, and only a very callous man would not have hesitated and hesitated again before giving orders as fateful as these.

The Germans, when they heard the news, were close to panic. Earlier that day, July 30, when he had learned of Russia's partial mobilization, the Kaiser's first anguished comment had been, "That means I have got to mobilize as well." The situation now was infinitely worse. The Russians were moving against them, the Germans, not just against the Austrians. And what could France do but support her Eastern ally? Here it was, the old nightmare of a two-front war, and it was real.

There was only one possibility, the Germans had long thought, of winning such a war. It lay in offsetting the overwhelming advantage in men and material that France and Russia held with speed. If they, the Germans, were to move quickly against the French and to defeat them in a matter of at most six weeks, they might then be able to rush their troops to the east just before the Russians had fully brought their forces into position. To allow the Russians to proceed with their general mobilization and thereby let Germany's only chance of victory slip away seemed intolerable. Or so the military experts in Berlin argued, and as in St. Petersburg, it was they who had the last word.

In the afternoon of Friday, July 31, two ultimatums went out from Berlin, one to Russia, the other to France. Would the Russians, stated the dispatch to St. Petersburg, suspend mobilization within twelve hours or face German mobilization and, by implication, war. Would the French, stated the dispatch to Paris, within eighteen hours give a promise of neutrality in the event of a Russo-German war. Should the French reply be affirmative—an unexpected contingency, since the ultimatum was no more meant to be accepted than Austria's ultimatum to Serbia had been—would they hand Germany the fortresses of Toul and Verdun in token of their good faith. Meanwhile, the Germans were preparing for the all too probable next step by proclaiming, that Friday, a "Threatening Danger of War," which meant that they were taking preliminary military measures just short of mobilization.

". . . And what would England do?" Both Haldane and Churchill had evaded a direct response to Ballin's question, but the answer, though still not final, was becoming fairly plain: Britain, too, was preparing to intervene. On July 31, Lord Grey asked both Germany and France for assurances that they would respect the neutrality of Belgium. The French quickly replied in the affirmative; the Germans were silent. The same day, Sir Eyre Crowe sent a long memorandum to Lord Grey. If, as seemed more likely than not, France and Germany should go to war, then Britain too should fight. "The theory," wrote Crowe,

> that England cannot engage in a big war means her abdication as an independent State. . . . The argument that there is no written bond binding us to France is strictly correct. There is no contractual obligation. But the *Entente* has been made, strengthened, put to the test and celebrated in a manner justifying the belief that a moral bond was being forged. The whole policy of the *Entente* can have no meaning if it does not signify that in a just quarrel England would stand by her friends. This honourable expectation has been raised. We cannot repudiate it without exposing our good name to grave criticism.

Almost imperceptibly, everyone's gaze shifted to Germany. The original causes of the crisis no longer seemed very important. On that same July 31, some eighteen hours after Russia's mobilization, Austria-Hungary ordered general mobilization. Hardly anyone paid any attention. To Russia and the *Entente* alike, Germany was becoming the main enemy. The Germans were paying the price for the distrust and friction that, through design or clumsiness, had been allowed to accumulate over the past two decades.

## War (August 1–4)

The events following Russia's mobilization and Germany's resort to ultimatums have something anti-climactic about them. The suicide pacts had been made; what was happening now, or so it seems in retrospect, was that the weapons were being raised, and raised with such resolution and speed that there was no chance for second thoughts.

On Saturday afternoon, August 1, the French, after rejecting the German ultimatum, ordered general mobilization. A quarter of an hour later, the Germans followed suit; there had been no reply from St. Petersburg to their ultimatum of the previous day. At about six o'clock that evening, the German ambassador, Count Pourtalès, called on Sazonov. Was there truly no hope of a favorable Russian reply to the German demand that mobilization cease? There was not, said Sazonov. "In that case," said Pourtalès, "I am instructed to hand you this note." He then gave Sazonov the German declaration of war on Russia, went to the window, and broke into tears.

It was up to the generals now. Their time to weep would come too; but at the moment, the generals, whether Russian, Austrian or German, wanted but one thing: to move! The day before Pourtalès's final visit, a significant scene had taken place in Vienna. In the midst of Bethmann's appeals to Berchtold to continue negotiating, a telegram had arrived from Count Moltke, chief of the German general staff, for his Austrian counterpart in Vienna, Count Conrad. Stand firm, said the telegram. Do not negotiate; "mobilize immediately against Russia." When Conrad showed the message to Berchtold, the foreign minister was taken aback. "Who is in charge?" he said. "Moltke or Bethmann?"

Moltke, of course. Ever since Russia's final decision to mobilize, the German generals had been impatient to begin military operations. They saw no other chance of victory except to strike first. They wished to defeat France rapidly and decisively while holding the defensive in the East and only then to shift their forces to attack the Russians. The question of course was how the Germans could manage to win such a victory over a France whose army was among Europe's best and whose border was heavily fortified. A plan, completed by the general staff under its then chief, General von Schlieffen, in 1905 provided a theoretical answer. The German army would circumvent the French fortifications by moving through Luxembourg and Belgium and take the French forces by surprise from the North. Would this violate Germany's treaty obligations toward Belgium? It would, but then, what was the alternative? Possibly, barely possibly, this was a justifiable counter-question for a professional soldier to raise. Possibly also, one should not be too astonished to find only the briefest references to the probable political repercussions of the violation of Luxembourg's neutrality ("no important consequences other than protests") in Schlieffen's memorandum of 1905 and none to those of Belgium's. What, however, of Germany's diplomats, who should have known better?

German diplomats, in the post-Bismarck epoch, did not normally care to argue with German generals. If that was what the gentlemen of the general staff held to be essential, said the gentlemen of the Wilhelmstrasse, then that was what German diplomacy must reconcile itself to. To quote the reported comment of Baron von Holstein, one of the senior officials of the German foreign ministry: "If the Chief of the General Staff, and particularly a strategic authority like Schlieffen, thought such a measure to be necessary, then it would be the duty of diplomacy to adjust itself to it and to prepare for it in every possible way." To which we might add "adjust, yes; prepare, no," since at no time between the plan's first draft and 1914 did the foreign ministry lay the least diplomatic groundwork for it.

With the decision to go to war against Russia, the general staff's Western war games became dead earnest. On Saturday, August 1, German troops invaded Luxembourg. On Sunday, August 2, at 7 P.M., the German envoy in Brussels presented an ultimatum to the Belgian foreign minister. (It was drafted mainly by Moltke; the foreign ministry had not prepared so much as the necessary diplomatic note in advance.) Germany, stated the note, to anticipate a French attack and "acting under a dictate of self-preservation," must march through Belgium. Should Belgium agree, the Germans would evacuate all Belgian territory upon the termination of the war and make restitution for any damage done. Should Belgium resist, "Germany would be obliged, to her regret, to regard the Kingdom as an enemy." Would the Belgians give an unequivocal answer within twelve hours.

This was worse than either a crime or a blunder. It was a disaster. The Germans were not only drawing the Belgians into the war against them; they were destroying all hopes of British neutrality. Quite possibly, England would sooner or later have joined Germany's enemies in any event, but owing to Berlin's lack of judgment, we will never know for certain. The violation of Luxembourg's neutrality had not disturbed the British unduly. (Here, for once, Schlieffen had been right.) It was true that Britain was signatory to a treaty guaranteeing the neutrality of Luxembourg, but the country was very small and possessed no coastline facing the channel. Belgium was different. The day before, the British cabinet had still been divided over what course of action to take. Pressures to intervene were coming from many sides: from the French, from the Conservative opposition in parliament, from the professionals in the foreign office. But the desire to stay aloof from the conflict (where really was Serbia?) was at least equally as strong. "I suppose a good three-fourths of our own party in the House of Commons are for absolute non-interference at any price," observed the prime minister to a friend as late as August 2. Hence the most that Grey had been willing to do had been to promise some naval support to the French in case the German fleet should sail into the channel to attack French shipping or the French coast. It was the ultimatum to Belgium that radically altered the situation. Britain, as a signatory to the Treaty of 1839, was one of the guarantors of Belgian neutrality (Prussia, and hence Germany, was another) and in 1914 as in 1839, Britain considered the defense of Belgium's channel ports all but a part of the defense of

PUNCH, OR THE LONDON CHARIVARI.—JULY 29, 1914.

THE POWER BEHIND.

AUSTRIA (*at the ultimatum stage*). "I DON'T QUITE LIKE HIS ATTITUDE. SOMEBODY MUST BE BACKING HIM."

*Blundering into War. The story of the war's origins—and of the much-discussed issue of responsibility—as much as anything is a story of missed opportunities and bungled moves. Thus in July 1914, Britain, whose role was as crucial as that of any of the powers, still tended to stay aloof from the conflict and even to display a certain understanding of the Austrian predicament.*

PUNCH, OR THE LONDON CHARIVARI. AUGUST 26, 1914.

THE TRIUMPH OF "CULTURE."

*Blundering into War: But by August, the German decision to invade Belgium had changed all that, and the demonization of the enemy had begun—on all sides.*

the realm. Later in the day of that same August 2 on which the prime minister had spoken of the desire for non-intervention, the British cabinet decided that if the Germans were indeed to violate Belgian neutrality, and if the Belgians were to resist, Britain would intervene.

Within forty-eight hours, there were no "if's" left. On Monday, August 3, the Belgians rejected the German ultimatum and called on the signatories of the treaty of 1839 for aid. That same day, the Germans began the invasion of Belgium and, on the mendacious pretext that the French had attacked German territory, declared war on France. On Tuesday, August 4, the British sent an ultimatum to Berlin. Would the Germans, within five hours, give an assurance that they were ceasing their attack on Belgium; if not, the British would sever diplomatic relations.

The Germans rejected the ultimatum. The quick defeat of France, they said, was a matter of life and death to them. One is struck by the simple-mindedness of the German response. What the Germans were doing, in effect, was asking the British to understand the requirements of the Schlieffen Plan, and then acting aggrieved when the British did not. The step that Britain was contemplating, said Bethmann, as the British ambassador paid his farewell call on him, "was terrible to a degree, just for a word 'neutrality,' a word which in war time had so often been disregarded—just for a scrap of paper, Great Britain was making war on a kindred nation who desired nothing better than to be friends with her."

Poor honest Bethmann, providing Allied propaganda with some of its most telling quotations. We have become somewhat less solicitous today over the universal observance of international law, but 1914 was another age. That same August 4, Bethmann rose in the Reichstag to say a few more words that, for his country's good, should forever have remained unspoken.

> Gentlemen, we are now in a state of self-defense, and necessity knows no law. Our troops have occupied Luxembourg, and perhaps have already entered Belgian territory.
>
> Gentlemen, that is a breach of international law. . . . The wrong—I speak openly—the wrong we thereby commit we will try to make good as soon as our military aims are attained. He who is menaced the way we are, and is fighting for his most precious possessions, can turn his mind only to how he might battle his way out.

At midnight, on August 4, Great Britain declared war on Germany. The Austro-Serbian conflict had escalated before the term had been invented.

*Even* Simplicissimus*—until then so strong in its criticism of its own society, and so muted in that of others—joined in, and would soon issue a selection of its anti-British cartoons under the title of "May God Punish England," with the money bag serving as Britain's new symbol. (Within less than six months, it had sold 50,000 copies.)*

Before it was over, the war would have expanded into a truly global conflict. The Central Powers were to gain only two more allies: Turkey, which entered the war in November 1914, and Bulgaria, which came in the next year. The Triple *Entente*, on the other hand, managed to draw in half the world on its side. Within the next three years, to mention only the major belligerents, Italy (which found new Allied promises more persuasive than old treaty commitments), Portugal, Rumania, Japan, the United States, and China were to join the Allied side.

The generals were in charge, and in August of 1914, a vast majority of people were deliriously happy that this should be so. The declarations of war seemed to have purified the air. In the various capitals of Europe, the crowds gathered—patriotic, fervent, elemental; athirst for victory and ready for sacrifice. "Whoever failed to see Paris this morning and yesterday," wrote Charles Péguy on August 3, "has seen nothing." He would not try to describe his journey to the front, wrote a French officer that same week; "sunshine cannot be reproduced adequately on a painting."

A few days earlier, after Berchtold had made his irrevocable decision on Serbia, the Bohemian writer Johannes Urzidil was walking through the streets of Prague. Coming to a tobacco shop, he noticed a group of people congregating around a telegram posted in the window:

> "War Declared"
>
> No one spoke. Then I suddenly heard three lone words: "*Gott sei Dank*"—"Thank God."
>
> The speaker was an Austrian first lieutenant. He did not try to give any reasons why one should especially thank God for this declaration of war. What will have become of him? . . . All I know is that the words "Thank God" soon changed into the much more fitting ones: "God help us!"

# CHAPTER 5

# A Question of Responsibility

## "Argument without End"

How had it happened? The victors, in 1919, were very certain of the answer—the Central Powers had willfully begun this most terrible of wars! The Germans, in Article 231 of the Treaty of Versailles, were made to admit their war guilt: "The Allied and Associated Governments affirm and Germany accepts the responsibility of Germany and her allies for causing all the loss and damage to which the Allied and Associated Governments and their nationals have been subjected as a consequence of the war imposed upon them by the aggression of Germany and her allies."

Thirty-three years later, a committee of French and German historians, engaged in an overall effort at removing some of the grosser errors of fact and judgment from both nation's textbooks, agreed on the following, rather different, statement: "The documents do not permit attributing, to any government or nation, a premeditated desire for European war in 1914. Distrust was at its highest, and leading groups were dominated by the thought that war was inevitable; everyone thought that the other side was contemplating aggression. . . ."

Between these two statements, tons of documents, books, and articles bearing on the cause of the First World War made their appearance. Their often very different presentations of what ought essentially to be the same facts are enough to shake anyone's faith in the feasibility of historical objectivity. And the argument is not over yet. Nor should it be. We all have good reason for wanting to discover how, particularly if no one really wished it, a world war of such proportions could occur. We also possess an advantage not of our own making: we should be far enough removed from the time, the place, and the passions of these events to reach some fair and sober judgments. What follows, then, is

**Confession**

*We hated war with all our heart*
*Thought nothing more infernal*
*But now we gladly play our part*
*To usher in peace eternal.*
Simplicissimus, *April 27, 1915.*

*The soldiers' uniforms are those of the Central Powers, but the illusion could be found on all sides.*

an attempt to do just that, although throughout, it will be well to bear in mind that splendid phrase of the great Dutch historian, Pieter Geyl, about "that argument without end which to us of the West is the study of history." Or, in the words of one of Theodor Fontane's Prussian squires: "Unassailable truths don't exist, or if they do, they are dull."

## National Responsibility

### *Austria and Serbia*

Let us begin by considering what, in descending order, each nation's share of the responsibility was. This is a consciously old-fashioned approach. The tendency, once the hatreds of the war had faded, was to emphasize less directly nation-centered causes of the war. But matters such as the alliance system, imperialism, national sovereignty, or the powers of the press were realities with which every statesman had to contend; they were a part of the atmosphere that affected all of Europe. How then did each nation manage its affairs in the real world in which it had to live and not in an ideal one of no entangling alliances, of colonial satiety, of the rule of world law, or of a press that promoted nothing but knowledge and love?

The nation that managed its affairs least well in 1914 was Austria-Hungary. We have stood stooped over too long now, searching for the underlying causes of the war. We have become so involved with subtleties that the obvious has sometimes escaped us—we have not seen the forest for the roots. The obvious fact is that it was the Austrian ultimatum of July twenty third that not only invited war with Serbia but laid the basis for general war as well. What an appalling document it was—tardy, incompetent, deceptive, designed to be rejected. Austria was setting the course, and neither friend nor foe had been allowed an honest look at its direction. Thus, to quote the leading expert on Austria's involvement, "it was Vienna," without any "pressure by Berlin to act," that "first resolved for war, that sought German assurances, and that exploited them once received."[1]

[1]Samuel R. Williamson, Jr. "Vienna and July 1914: The Origins of the Great War Once More." In Williamson and Peter Pastor, eds., *Essays on World War I: Origins and Prisoners of War* (New York: 1983, p. 24).

It is entirely possible, of course, to present a case for the defense. It takes no major effort to understand why those who set policy in Vienna were convinced that this time Belgrade had gone too far, and that hence "only militant diplomacy remains an option," or as the foreign minister put it even more forcefully, that the time had come for a "final and fundamental reckoning" with Serbia.[2] Sarajevo, after all, was more than a single incident. As seen from Vienna, many another provocation had preceded it. And the Habsburg monarchy had some perfectly legitimate interests to defend—Serbia was immensely closer to Austria than the many regions of the world that Washington has by now considered to be vital to its interests are to the United States. Surely, Austria's desire for self-preservation was as sensible and honorable a motive as any. Besides, the state that one wished to maintain had a great deal to be said in its favor. "My whole libido," wrote Dr. Freud the day after Serbia's reply had been received, "goes out to Austria-Hungary."

But to understand all is *not* to forgive all. It was true enough that no government or nation had "a premeditated desire for European war in 1914." But Austria, in 1914, had very much of a premeditated desire for a small Balkan war. Did it want that war to spread? No, but the truth was that the people who were deciding policy in Vienna did not really care. This was Austria's war; perhaps only the fact that the countrymen of Johann Strauss and Sigmund Freud ordinarily made such poor villains (there would be some fairly obvious exceptions such as Adolf Hitler) allowed that simple fact to be forgotten so thoroughly.

Austria's war, and Serbia's. Colonel Dimitrijević, too, had not much cared what the repercussions of Franz Ferdinand's murder might be. Had he known that his deed would mean war, said one of the assassins during his trial, he would have preferred his own death to that of his victim. It is not likely that he was speaking for Dimitrijević. But to what extent did the action of Serbia's chief of army intelligence incriminate his government? The murder, after all, had been planned by the Black Hand during the colonel's off-duty hours and not by Pašić or any of his ministers. The answer is that Dimitrijević's action did incriminate the government, for reasons both long-range and immediate. For one thing, it is difficult to take exception to the observation of a close student

[2]Samuel R. Williamson, Jr., *Austria-Hungary and the Origins of the First World War* (New York, 1991: pp. 194 and 213).

of the period that "the fact that the Serb government could not control its own army or nationalists only in part diminishes its responsibility for the murder."[3] Beyond that, what had happened on June 28 in Sarajevo was hardly all that astonishing, since that government had for too many years been tolerating or even encouraging a movement for a Greater Serbia, whose aims were bound to be offensive to Austria-Hungary, and whose methods were bound to be offensive to anyone. Very specifically speaking, the government in 1914 had taken no effective action to prevent the assassination of Franz Ferdinand, of which it very probably had some foreknowledge, nor had it managed to end the influence of the Black Hand, of which it assuredly had knowledge.

Again, there was nothing ignoble about either Pašić's or Dimitrijević's aims. The concept of a Greater South Slav State was fully as defensible as was Austria-Hungary's right to survival. Tragedy, in the Hegelian definition, consists not of the conflict of right with wrong but of right with right. But the Serbians set about achieving their purposes with a truly frightening disregard of the consequences. Here, then, was Serbia's vast share in the responsibility for the First World War, one that was matched only by Austria's: Belgrade surely knew that it was set on a collision course, yet it would not alter direction. There is, in the British files, a report from the ambassador to Vienna, Sir Fairfax Cartwright, written in January 1913 in connection with another, milder crisis brought about by Pan-Serb agitation, which sums up the entire matter better than any later historian can:

> Serbia will some day set Europe by the ears, and bring about a universal war on the Continent. . . . I cannot tell you how exasperated people are getting here at the continual worry which that little country causes to Austria under encouragement from Russia. . . . It will be lucky if Europe succeeds in avoiding war on account of the present crisis. The next time a Serbian crisis arises, I feel sure that Austria-Hungary will refuse to admit of any Russian interference in the dispute and that she will proceed to settle her difference with her little neighbor by herself *coûte que coûte*.

## *Russia and Germany*

Russian encouragement and Russian interference—one wonders how different Serbia's course of action might have been in 1914 without

[3]D. C. B. Lieven, *Russia and the Origins of the First World War* (New York: 1983, p. 139).

these. In the sixty years that preceded Sarajevo, Russia had been involved directly in two wars—the Crimean War and the War of 1877–1878—indirectly in two more—the Balkan Wars of 1912 and 1913—and, directly or indirectly, in crises too numerous to list, all with the intent of weakening the empire of the Sultan to the benefit of that of the Tsar. It was a policy that, with its implications for the Balkans, acted as an irritant to Austro-Russian relations; threats to one's survival tend to annoy. Yet the Russians persisted. It was Russia rather than Austria that was the expansionist power in Southeastern Europe. The Annexation Crisis was the exception, not the rule. Under ordinary circumstances, Austria had considerably more to fear from Russia's Balkan ambitions than Russia had from Austria's, even if we add the reservation that, customarily, Russian official policy was more restrained and rational than some of the truly extravagant Pan-Slav spokesmen would have liked. Customarily, for in July 1914, as it had on some previous occasions, the borderline between Pan-Slavism and government policy once more became blurred. If Serbia chose to resist Austria's demands, then Russia would support its Slavic brethren. The story of Germany's "blank check" to Austria is well known, but the Russians were doing no better at counseling caution to Belgrade. Yet just as Germany's decision to support Austria was a crucial element in the crisis leading to war, so was Russia's decision to support Serbia. Or to quote one of the most knowledgeable of experts on Russia's role in 1914, Dominic Lieven, "Petersburg's stand made a European conflict probable."

Perhaps St. Petersburg had no real alternatives. Caution might be seen to equal surrender, and to quote Lieven again, the Russian establishment, for all the "lack of coordination or even mutual comprehension between the State's diplomatic, military, and political leaders . . . would not accept a clear and insulting demotion to the ranks of Europe's second class powers."

What made that concern all the more palpable was the memory of 1908. If Russia should once again fail to intervene, the damage inflicted on Russian prestige in Serbia and beyond would be incalculable. To abandon Serbia now, said the Russian foreign minister in the cabinet meeting of July 24, 1914, which in essence set the signal for war, would mean the betrayal of Russia's "historic mission" as the protector of the Balkan Slavs, and relegate the country to "second place among the powers."

The record of that meeting was confidential of course, yet the gist of what he was saying was known well enough. The Austrians, certainly, had no reason to assume that Russia would merely look on, let alone look the other way, while they were shooting their way into Belgrade.

Yet the Russian response, it should be noted, consisted of more than a move to halt the Austrians. Had Russia mobilized its forces against Austria only, a negotiated peace, taking both Russia's and Austria's legitimate interests into account, might still have been possible. The Russians did not do this. Instead, and here we come to Russia's quite staggering responsibility for World War I, they decided to marshal their forces against Germany as well as against Austria. Their reasons for general mobilization were technical rather than political; honest error and the absence of flexible military planning rather than premeditated design accounted for the Tsar's final decision on July 30. But no amount of explaining can change two facts. One was that Russia, in 1914, was the first power to mobilize. The other was that the Tsar and his advisers, by their behavior on July 29 and 30, showed that they were perfectly well aware of what mobilization meant.

What, though, of Germany's guilt? Was the Versailles indictment really that absurd? Surely, it could not have been due entirely to the machinations of Allied propaganda that Germany was seen as the chief culprit in 1914 and not Russia, Austria, or Serbia, or that so many subsequent accounts (this one included) written in a more temperate climate of opinion, would place Germany so close to the center of the narrative.

German responsibility for 1914 goes back at least to the 1890s, to the whole conduct of German diplomacy after Bismarck. The master's touch was shown by moderation, the touch of his successors by what Bismarck had accused post-Cavour Italy of: a large appetite and poor teeth. And to what did Germany's appetite really run after 1890? To "world power," to rivalry with Britain, to a German Central Africa, to security against France? None of these policies had been thought through. Activity there was aplenty; what was lacking was direction, and so was a decent regard for the concerns of others. No one put it more sharply than the man who had the misfortune to hold Bismarck's office in 1914, Theobald von Bethmann Hollweg. "We have lived a lie," he said during the war's first year, "both in our domestic and our foreign policy. A spirit of the raised voice, of over-impetuousness, of loquaciousness, and of showing off has come to infest our people."

Thus by 1914, a new specter was haunting Europe. The specter of *furor teutonicus*. German behavior in the final crisis certainly did little to banish it. The list of German blunders is an all too familiar one: the blank check to Austria; the failure, once it had become plain that an Austro-Serbian war was likely to become European in scope, to search for a compromise with every last ounce of energy; the final triumph of the military over the political rationale with the invasion of Belgium and France. Bethmann himself again summed it up best perhaps in a conversation with a visitor to his country estate toward the end of the war. Germany, he said, and the emphasis was his, had pursued "a **policy of utmost risk,**" and thus "bears a large part of the blame."

The extent of that German blame has become so familiar by now that one is tempted to point out that there is another side to it. One is especially tempted to do so in the light of the discussion initiated by the Hamburg historian Fritz Fischer. Fischer has maintained, in one of the most stimulating of reinterpretations of modern diplomatic history, that Germany's leadership had a will to war in 1914, and it was Germany, with its "bid for world power" and its readiness to engage in conflict with Russia and France, that bore a crucial part of the responsibility for the castastrophe.

Now it is disarming, and probably very healthy beside, to have a historian take this critical a look at his own nation's recent history. What is wrong with it, however, and with general allegations of Germany's war guilt as well, is that it involves a certain amount of confusion between Germany's war aims as they ultimately developed and Germany's intentions in July 1914. What is wrong with it, too, is that it depends for its effect on viewing Germany's actions in isolation. Germany indeed wished to be a great power, and why not? The concept of a German Samoa was no more absurd than that of a British Bahamas or a Belgian Congo.

But Germany, the distinction may be worth spelling out, wished to be a world power; it did not aim at world domination. To be sure, during the course of the war, some fairly grotesque German war aims emerged, from the annexation of Belgian iron mines to the establishment of a German Central Africa. But two things need to be noted. One is that Allied war aims were hardly more restrained. The Russians, French, and British were equally busy composing memoranda (and in some cases binding treaties) on the dismemberment of Austria, the truncation of Germany, the transfer of the Dardanelles to Russia, the acquisition of vast tracts of the Ottoman Empire by Britain and France.

The other is that in August 1914, Germany no more went to war for Belgian iron ore than Russia did for the Straits or Britain for Palestine. That appetite grew with the eating. Motives are seldom simple, of course, and one might say that expansionist thoughts were in the back of some statesmen's minds that summer, but that was exactly where they were, not in front. In the mountains of documents on the origin of World War I, we search in vain for any that will show us that the statesmen of Germany or of any other nation (with the possible exception of Serbia) went to war in 1914 for the sake of their later war aims.

None of this absolves the Germans from their share in the responsibility. But it is a share only, to be divided with at least five other powers. Germany's responsibility, to reiterate the point, lies in the general foolishness with which the nation behaved in the quarter century before Sarajevo and, more immediately, in the panicky and pernicious decisions that Berlin made during the July crisis. These things were bad enough; there is no need to add imaginary to real guilt.

## *Britain and France*

Was it really five powers with which the Germans shared the responsibility? That Germany could point to Russia's mobilization, that Russia in turn to Austria's threat against Serbia, and Austria to Serbia's initial provocation should be obvious, but were not Britain and France innocent bystanders? Not quite.

England's responsibility, clearer than that of France, lay not in the allegedly insufficient warning the British gave the Germans of their intent to fight, the topic that was so widely discussed in the years immediately after 1914. Surely, any responsible German statesman should have known that it was a matter of vital interest to Great Britain whether France survived as a power and who would control the channel ports of Belgium and France. The implications of the *Entente Cordiale*, and the very plain provisions of the Treaty of 1839, were quite sufficient warning here.

What was involved in the case of Britain was something more long-range. There had taken place, in the half century or so before the war, a tremendous expansion of British power, accompanied by a pronounced lack of sympathy for any similar ambitions on the part of other nations. If any nation was compensation-conscious, it was Great Britain; if the Austrians wished to occupy Bosnia, for instance, then the British must have Cyprus. Even without this particular diplomatic gambit, the British, between the eighteen-seventies and the turn of the century, were adding, adding, adding to their Empire: Burma, Egypt, Uganda, Somaliland, Kenya, Zanzibar, Rhodesia, the Boer Republic, all were flying the Union Jack. If any nation had truly made a bid for world power, it was Great Britain. In fact, it had more than bid for it. It had achieved it. The Germans were merely talking about building a railway to Baghdad. The Queen of England *was* Empress of India. If any nation had upset the world's balance of power, it was Great Britain.

The usually more modest appetites of others, however, were a different matter. That the French in Morocco or the Germans in South West Africa were doing quite as well as colonizers as were the British in Uganda was something that seemed all but impossible to perceive through the London fog. British territorial acquisitions were a part of the progress of mankind; those of others were a menace to world peace and civilization. Russia's desire for access to the Mediterranean was a provocation; Britannia Rule the Waves reflected a noble and natural sentiment.

After the turn of the century, this attitude was directed with special force at Germany, although it was seldom made clear just what the Germans were to do with all *their* excess energy. Some of the British documents of the period convey an impression of near-hysteria in the face of a rising Germany, reaching from a British admiral's simple suggestion, cited before, to sink the German fleet without warning, to the more erudite memoranda of some of the senior officials of the foreign office decrying the German menace. Unfortunately, if one nation imagines for long enough that another is a menace—"the natural enemy" of Sir Eyre Crowe's phrase—the likelihood is strong that that other nation will some day have to play the role it has been assigned.

That judgment does not exculpate the Germans, who had been willing to endorse a Balkan war that clearly contained the risk of a war in the West. As far as Great Britain's direct responsibility for the final crisis and the outbreak of hostilities is concerned, it is indisputably less than Germany's. It was units of the German army, not of the British navy, that were lobbing shells into Belgian towns. Before the guns had been moved into position, however, it can scarcely be said that the British did much better in restraining the Eastern member of the Triple *Entente* than they would accuse the Germans of doing vis-à-vis Vienna. "Will to war" is too grand and at the same time elusive a term to use with any degree of comfort, but what do we do about the memory of what would happen a quarter of a century later? In 1938, when the British definitely did not wish war, their prime minister would hurry to Germany three times within the space of two weeks and negotiate peace under terms far worse than those of 1914 with a leader infinitely more dangerous than the Kaiser in his most Wagnerian moods. The lamps were not "going out" (Lord Grey's celebrated figure of speech) in 1914. Many people, the author of the phrase included, were helping to put them out.

Last of all, there is the responsibility of France—last of all, and least of all. Had the ultimate decision been that of France, the lamps would have remained lit.

It was true that a generation or more after Frankfurt, the French still had not really reconciled themselves to the loss of Alsace and Lorraine. Poincaré and his friends were still pursuing what Boulanger had, except that they were doing so with greater intelligence, tact, patience, and skill and hence with greater effectiveness. The desire to undo the decision of Frankfurt was as powerful a motive force in France's system of alliances as any. Germany, to put the matter differently, did not wish for any territorial change in Europe between 1871 and 1914; France did. It was equally true that during the July crisis, France failed to urge restraint on St. Petersburg. One cannot very well indict the Germans for their blank check to Austria without noting that the Russians held a similar piece of paper from France.

Yet when compared with the involvement of other nations, how mild the responsibility of France must appear. The French might mourn Strasbourg, but they were plotting no war to recover it. Even the most irreconcilable groups of Frenchmen were organizing no Pan-Gallic movements; there was nothing in France to compare, for ambition and folly, with either the Pan-Slavs or the Pan-Germans. Also, all speculations about France's ultimate objectives have about them a certain irrelevancy. Perhaps France would, some day, have fought to avenge Sedan, perhaps France would not. We will never know. The fact is that the French, in August 1914, did not go to war for Alsace, but because the Germans, lacking the political imagination and the diplomatic skill to keep France neutral, first presented an unacceptable ultimatum and then began to march on Paris. The British had an element of choice, even after the Schlieffen Plan had gone into operation (major segments of British opinion certainly thought so); the French did not.

In 1914, the French entered the war because they had no alternative. The Germans had attacked them. It was that simple.

## Other Causes, Other Reasons: Diplomacy to Public Opinion

Were not certain other factors more complex, however? What of the forces which transcended traditional national responsibility: the alliance system, the role of economic motives, or the impact of imperialism, nationalism, and the press?

Having dealt with these topics before, what can be said rather summarily here is that while these factors unquestionably affected the atmosphere of 1914, and usually for the worse, none of them, by themselves, were enough to cause a war.

The alliance system, with its resulting division of Europe into two camps, did contribute to transforming Austria's war into general war. But the alliances were effect, not cause. They were the traditional and legitimate instruments of each nation's often equally legitimate and traditional objectives, and not ends in themselves. Nor were the alliances nearly so firm and automatic as they sometimes appear in historical literature. They certainly were considerably less integrated than the system of alliances, both East and West, that emerged after the Second World War. Italy, for instance, would remain neutral in 1914 despite its formal treaty obligations to the Central Powers. The Ottoman Empire, on the other hand, would join them even in the absence of any such ties. In fact, one might well advance the proposition that if only the alliances had been stronger—if only Austria-Hungary had been absolutely certain that an attack on Belgrade meant an attack on St. Petersburg, if only there had been no doubt at all in the Germans' minds that a move against Brussels was a move against London—everything might still have ended, if not well, at least short of disaster.

There is even less strength in the economic argument, so tempting to Marxist historians, who professed to believe that "the anarchy of the capitalist world" was bound to lead to bloody conflict. And it was true, of course, that the economic rivalry between the powers, most notably between Germany and Great Britain, did not contribute to international amity. But few were the businessmen who would risk the destruction of their entire national economy for the sake of increasing their share in the export market. Businessmen were altogether capable of seeing the advantages of peace, and of economic cooperation. In the case of the Berlin to Baghdad Railway scheme, for instance, British and French bankers were very much in favor of participating along with their German confreres; it was their governments that persuaded them to abstain. And in 1914, businessmen, particularly in Britain and Germany, were as vociferous as anyone, workers included, in opposing war.

Nor should the influence of imperialism or even of the press be exaggerated. Conflicting colonial ambitions did add to Europe's ill temper, especially where Anglo-German relations were concerned. But then, they were an equal burden on Anglo-Russian and Anglo-French relations; yet in 1914, England, Russia, and France found it perfectly possible to fight on the same side. Besides, as a leading expert on British policy, Zara S. Steiner, has written "Rather than a cause of war, the division of far-off lands had often provided a safety valve for adjustments that preserved the European peace."

The press, with its tendency to exacerbate or invent diplomatic crises, with its daily offerings of international disaster (potential, imagined, or real), and with its habitual blurring of the distinction between patriotism and chauvinism, may have more to answer for. But public opinion alone can create no wars. If it could, the French would have gone to war against the British at the time of Fashoda and the Serbians and Russians against the Austrians at the time of the annexation crisis. Even William Randolph Hearst, once we separate fact from legend, was not able to deliver a Spanish-American War quite by himself.

## The Matter of Militarism

One topic that cannot be dealt with quite so cavalierly is the question of militarism. Militarism is a vague and emotional term, so let it be said that two fairly specific aspects of it will be of concern here: the amount of arms available in 1914, and the more than occasional preponderance of military over political considerations.

The military expenditures of the European powers, as noted before, rose at an unprecedented rate in this period, until, at the beginning of 1914, they stood at 300 percent of what they had been in 1870. While the amount of weapons was thus very clearly measurable, the influence on policy was a more subtle matter. Yet Hegel's observation that at some point quantity will change quality was as true as ever. There were too many deadly weapons about for their owners' good in 1914 Europe. With all these splendid instruments of destruction available, it would have been surprising if there had not been a few people who had no marked aversion to seeing them tested.

Their numbers were small, but Austria's Berchtold was too close to them and so was Russia's Janushkevich. This brings us to the second and even more critical aspect of the matter, the undue influence of military over political decision making. The Germans and the Russians were the worst, though not the only, sinners here. Instead of serving policy, the preparations undertaken by Russia's and Germany's military authorities—general mobilization and Schlieffen Plan—decided policy and, as it turned out, disastrously so.

Relatively little blame attaches to the generals. It is their task to prepare for war and victory. Nor were the generals always automatically members of the war party. In January 1914, for instance, when Sazonov had advocated the forcible seizure of the Straits, it was the Russian military experts who had opposed the venture on the grounds of its impracticability. A great deal of blame, on the other hand, attaches to the statesmen. It was they who should have known how stretchable a term "military necessity" was, who should have seen how perilous the plans of their respective general staffs were and have countermanded them.

Why they did not is not an entirely easy question to answer. One can of course say the obvious, which is that they were men who were not given the qualities necessary to establish the absolute priority of civilian over military authority, qualities such as toughness, great clarity of purpose, and the ability to keep regular office hours. (Berchtold and Sazonov, one suspects, were so much the prisoners of their generals' plans in July 1914 largely because they had never bothered to find out just what those plans were.) And one can add that it was in those countries where the military stood at the top of the social hierarchy—attracting talent and bestowing prestige as perhaps no other profession did—that there occurred the worst violation of the rule, essential to sound government and sane diplomacy alike, that the sovereign must be commander-in-chief, not the commander-in-chief sovereign.

But this is not the whole answer. For in Britain and France, too, membership in the general staff tended to be an indication of more than usual intelligence and social acceptability. Nor did the military staff talks held under the shelter of the *Entente Cordiale* invariably follow the instructions of the representatives of people or crown; in the summer of 1914, the majority of the British cabinet would be quite as surprised by the obligations incurred toward France as was the majority of the German government by the modalities of the Schlieffen Plan. Still, there were some rather important differences in degree. Thus the French government, when presented with its counterpart of the Schlieffen Plan, providing for a French invasion of Belgium, told the general staff to discard it and to submit a politically more tenable one. And while the professional soldiers of all nations were perhaps equally fascinated by the technological necessities of modern warfare, it was in Russia and Germany again that technology was allowed its clearest triumph over diplomacy.

World War I, as Herman Kahn has been among the first to point out, was to a large extent a war of railroad timetables: "the general staffs of the four great continental powers had spent decades planning meticulous timetables. The war plans were literally cast in concrete in the sense that governments built railroads according to the requirements of the war plan."[4] Now the advantages of railroads, demonstrated on a large scale first in the American Civil War and then in the Franco-Prussian War, were considerable. They could be used to deploy large armies quickly, effectively, and cheaply and to transport these armies to fight on territory other than one's own. But the system also was notably rigid. A horse can change direction at the flick of the bridle, not so a train. Once set in motion, how was it to be stopped? Even if stopped, how could it be redirected without creating havoc with the remainder of the schedule? Hence, in the Russo-German impasse over mobilization, very much more was involved than diplomatic precedent or prestige. The means of war were being allowed to bring about a state of war.

The final irony was that as the armies began to move, not one of the carefully calculated timetables would really work. Everyone's plans had been predicated on what one historian has called "the short war illusion," that is to say on swift and surprising movement, on quick and decisive victory—or on precisely the sort of thing that would not come to pass. What no one had anticipated, to make a long mistake very short, was that the war they were about to fight was World War I.

## "A Tremendous Lack of Imagination"

That was hardly the only miscalculation. Everyone was so intrigued by what might happen if they did *not* risk war that they failed to perceive

[4]Herman Kahn, *On Thermonuclear War* (Princeton, 1961), p. 359.

the cost of violence. The alternatives to war, in the innocent summer of 1914, thus seemed very much worse than war itself. The Austrians were mortally afraid that if they did not march, they might forfeit the final opportunity to save their empire. The Russians felt that they could not possibly leave their Serbian allies to fend for themselves again as they had during the annexation crisis. The Germans concentrated their fear and imagination on the consequences of losing their one remaining ally. The French were convinced that their alliance system might suffer an irreparable blow if they as much as hesitated in aiding Russia. The British did not see how they could acquiesce in German control of the channel ports, let alone in German dominance over Europe.

None of these fears were groundless, but war was about as good an answer to them as suicide is to the fear of death. Hence what strikes one, at least at this distance in time, is that there was no all-out search for an alternative, but that instead the crisis was allowed to run its more or less routine diplomatic course. The price for that would be horrendous. The war's direct cost, often calculated and described, still fits no category of horror that we are capable of visualizing. At Verdun alone, perhaps a million Frenchmen and Germans died. How does one picture a million deaths? How does this small book hold the bloody fields of France? And bloodied to what end? To the end that the war would destroy the Austro-Hungarian Empire it was designed to save, end Russian influence in the Balkans for years to come, and leave Germany more weak and lonely than she had ever been before. Nor would it spare the victors. Without it, Britain might still be the world's banker and France still the master of an empire extending from Algeria to Indo-China. "This war," said Franz Kafka to a friend in 1916, "above all else was caused by a tremendous lack of imagination."

The lack was understandable. Imagination, in cases other than Kafka's, more often than not is the extension of the past into the future, and the past was reassuring. None of the many crises to which people had been witness *had* led to war. Nations had threatened one another, had dispatched curt notes and large dreadnoughts, but at the final moment, they had always drawn back, whether over Fashoda, Morocco, or Bosnia, since none among them really wanted a general war. Hence, what each side expected was the customary mixture of public bluster and private good sense. Once more, there would be the walk to the brink; once more, one would, with a suitable show of resistance, permit oneself to be pulled back.

Overconfidence, the alliance system, the press, national antagonisms, militarism—the list still is incomplete unless there is added to

Alarm

„Zu den Waffen! Ihre Interessen in Marokko sind bedroht!"

*"A Tremendous Lack of Imagination": "Alarm" is the heading of this classic cartoon by Eduard Thöny in* Simplicissimus *for June 1, 1908, whose caption reads, "To Arms! Your interests in Morocco are threatened!" From this distance in time, it is tempting to substitute Bosnia for Morocco, and to let it equally stand for what was being allowed to happen in 1914.*

it the crucial event that made these and other factors operable. That event was the crime of Gavrilo Princip. Without Sarajevo, Europe might have remained at peace for no one can say how many years to come. 1914 was not 1939. Too much was right with Europe then. The world had weathered (and would again weather) considerably worse periods of crisis without recourse to global conflict. Or as a poem by Arthur Guiterman, written on the occasion of an earlier crisis, that of the Annexation of Bosnia, and published in the *New York Times* of October 9, 1908, had it:

> How I love to watch the War Cloud, as it
> gets an extra head on!
> How I love to hear the "trouble,
> trouble, trouble!" of the drums,
> And the wailings of the Prophets for the
> Day of Armageddon;
> It's so thrilling and romantic—and the
> trouble never comes!
>
> Though the Powers are pow-wowing and
> the Kaisers are contriving,
> Yet we know one party dassen't and the
> other one's afraid.
> There'll be scores of ultimatums and a bit
> of bargain driving,
> And then all will be as peaceful as a Tem-
> perance Parade.

A dubious forecast, as it would turn out, but then history is not prophet-friendly. Attitudes, alliances, conditions, constellations of power all have a tendency to change in the normal course of events, and often rapidly and unpredictably so. How they had changed between the 1880s and the turn of the century, for instance, and how, with Franz Ferdinand alive, they might have changed once more!

Sarajevo, then, was a major cause of the war, not merely its occasion. Princip killed more than two victims. It was his and Dimitrijević's deed, as well as the errors of fact and judgment that occurred in its aftermath that gave all the various elements of friction their unique opportunity of coalescing. The errors and the pure bad luck, the sheer accidents, from a chauffeur's wrong turn to a German foreign minister's absence honeymooning to a Russian chief of the general staff's being new to his office.

No power had any overall, conscious designs for war in 1914. Nowhere, even in the summer of 1914, was a calculated, advance decision made for global war. Rather, the powers, as a result of Sarajevo, became involved in a series of moves and countermoves—all of them, on the face of it, logical, reasonable, or at least defensible—that stage by stage, step by step, imperceptibly at times, and hardly ever with any true vision of the consequences, placed them in a position from which there was no way back to the negotiating table. Thus, all the trends, events, factors mentioned in these pages, both long-range and immediate, had their bearing on the origins of the First World War and were in some manner relevant. Thus, too, none of them were. For not one of the factors described was sufficient in itself to cause a war, nor had, in previous years, at least partly similar combinations of circumstances ended the way Sarajevo did. Perhaps all one can truly say in the end is that World War I was a twentieth-century diplomatic crisis gone wrong, the one gamble, or rather series of gambles (of how many in all, before or after?) that did not work out. Such things happen.

# Bibliographical Note

All these many years later, three books still form the core of the bibliography of the Causes of the First World War: Sidney B. Fay, *The Origins of the World War* (Macmillan: New York, 1930), Bernadotte E. Schmitt, *The Coming of the War, 1914* (C. Scribner's Sons: New York, 1930), and Luigi Albertini, *The Origins of the War of 1914* (Oxford University Press: London, 1952–1957). Some of the details in Fay and Schmitt are now outdated, but on the whole, both books have stood the test of time remarkably well. There has been a Library of Congress card explosion of World War I titles since the appearance of Fay and Schmitt; there have been more primary sources published than even teams of historians could absorb; there has been an entire scholarly periodical that, for the two decades of its existence, devoted itself to nothing but the question of the causes of 1914. Throughout it all, Fay and Schmitt have remained indispensable. Even their biases—Fay tends to absolve the Central Powers of the Versailles war-guilt accusation, while Schmitt holds it to be essentially true—have not really detracted from their worth. Rather, both have defined and documented scholarly positions that all subsequent historians have had to take into account.

The Albertini book is in a somewhat different category. It is a book more likely to be quoted than read. It is a jungle of detail, in which supposition and fact, the germane and the inapposite, the new discovery and the old cliché, the marginal and the essential all dwell side by side, with few discernible efforts on Albertini's part to provide paths or clearings. But if it is a jungle, it also contains treasure. Scarcely a document of any bearing on the topic has escaped the author's scrutiny. (Many of these, of course, were not yet available when Fay and Schmitt appeared.) Hardly an event remains undescribed. In addition, he interviewed or corresponded with many of the participants in the events, asking some shrewd and probing questions and receiving some revealing answers. In sum, then, Albertini's three long, complete, and sometimes exasperating volumes stand as a reference work that any serious student of 1914 will have to consult.

There are, in addition to these three, literally thousands of other titles that could be mentioned. Nearly all of the governments involved have released the relevant diplomatic documents, in collections that often go back to the latter part of the nineteenth century. (The principal collection of French documents alone runs to three series and over thirty volumes, and there are, in this case as in others, several supplementary collections as well.) In addition, most of the statesmen active at the time have written their memoirs, and there are numerous biographies of both major and minor political figures that in many instances are based on personal or official documents not available elsewhere. There also exists a vast secondary literature on the topic; in quantity of material, World War I is a serious contender to either Napoleon or the American Civil War.

Fortunately, there is no need to attempt the provision of even a skeleton bibliography here—which in its leanest form would still require a good many pages—since several excellent such bibliographies are easily available. Three of the best are the bibliography in volume III of Albertini; the bibliographical essay in Dwight E. Lee, *The Outbreak of the War, Causes and Responsibilities*, Fourth Edition (D.C. Heath and Company: Lexington, 1975); or the "Suggestions for Additional Reading" in Holger H. Herwig, ed., *The Outbreak of World War I, Causes and Responsibilities*, Fifth Edition (D.C. Heath and Company: Lexington, 1991). The last two titles mentioned, with their selections of writings on the topic, also offer fine introductions to the sometimes radically different interpretations of the war's causes offered by various historians whose scholarly distinction has proved no bar to reaching opposite conclusions based on what often is the same evidence.

As for another approach to the topic, that of the biographies of the major actors in 1914—particularly on the German side—and of the years leading up to that fateful date, just a few titles might be mentioned. It is a list which, because of its potential length, is perhaps arbitrary (and certainly not complete) even though here as elsewhere it will be restricted to titles available in English. Among the biographies of William II, three in particular remain worth citing. They are Michael Balfour, *The Kaiser and His Times* (Houghton, Mifflin: New York, 1972), Lamar Cecil, *Wilhelm II* (University of North Carolina Press: Chapel Hill, 1989), and Virginia Cowles, *The Kaiser* (Harper & Row: New York, 1963).

Among the biographies of the Chancellor dismissed by William II to his sorrow, two of the most thoughtful, recent, and readable are Edward

Crankshaw, *Bismarck* (Viking Press: New York, 1981, paper 1983), and Lothar Gall, *Bismarck, The White Revolutionary* (Unwyn Hyman: Winchester, MD, 1990). Another, by Hans-Ulrich Wehler, an expert on period and subject, remains untranslated, but we do have his survey on *The German Empire, 1871–1918* (Berg Publishers: Dover, NH, 1985), which contains much that is of interest in connection with the Chancellor and his times. The same is true of an American historian, Gordon A. Craig, whose *From Bismarck to Adenauer, Aspects of German Statecraft* (Johns Hopkins Press: Baltimore, 1958) offers a perceptive and still very much valid analysis of Bismarck's diplomatic style. But outstanding though Craig remains, perhaps the first book to which a reader might want to turn is Sebastian Haffner's brilliant *The Ailing Empire, Germany from Bismarck to Hitler* (Fromm International: New York, 1989), which gives Bismarck the credit due to him for championing, in his later years, "the only broad-based peace policy ever pursued by the German Reich." As for the man who had the bad luck of occupying Bismarck's office in 1914, we have the fine biography by Konrad Jarausch of *The Enigmatic Chancellor, Bethmann Hollweg and the Hubris of Imperial Germany* (Yale University Press: New Haven, 1973). (The source of the quotation cited on p. 137 about "the policy of utmost risk" pursued by Germany in 1914 comes from the same author's article "The Illusion of Limited War: Chancellor Bethmann Hollweg's Calculated Risk, July 1914" in *Central European History*, II, 1, March 1969, p. 48.)

For the view from the other side of the channel, some outstanding titles are Paul M. Kennedy, *The Rise and Fall of the Anglo-German Antagonism* (Allen & Unwin: London, 1980), and Zara S. Steiner's *The Foreign Office and Foreign Policy* (Cambridge University Press: Cambridge, 1969), which offers some of the sharpest closeups that we have of the inner workings of the British diplomatic establishments, as well as her *Britain and the Origins of the First World War* (St. Martin's Press: New York 1979), which relates even more directly to our topic. (The quote on the preference for "absolute non-interference" on p. 129 is taken from *Britain and the Origins*, p. 232.)

On Russia's role, some excellent work is at long last available. Two authors among all others deserve to be mentioned here. One is D. C. B. Lieven, whose *Russia and the Origins of the First World War* (St. Martin's Press: New York, 1983) is particularly good at showing that it was not domestic pressures, as sometimes claimed, that made Russia behave as it did in 1914, but rather the fear—similar to Germany's and Britain's—that to act other than it did would jeopardize its great-power status.

(The passages quoted on p. 135 appear in Lieven's book on pp. 139 and 153–154.) The other writer is George F. Kennan, diplomat and scholar who has yet to write a bad book. Applicable titles are *The Decline of Bismarck's European Order, Franco-Russian Relations 1875–1890* (Princeton University Press: Princeton, 1979), and *The Fateful Alliance: France, Russia, and the Coming of the First World War* (Pantheon Books: New York, 1984). (The quotation on the Franco-Russian Military Convention on p. 29 is from his article, "The Balkan Crisis 1913 and 1993," in *The New York Review of Books*, July 15, 1993, p. 3.)

For the part played by Austria-Hungary, a good many years had to pass before the definitive research on the topic found its way into print. But we now have it with Samuel R. Williamson, Jr.'s *Austria-Hungary and the Origins of the First World War* (St. Martin's Press: New York, 1991), as well as in the same author's equally fair and concise "Vienna and July 1914: The Origins of the Great War Once More," in *Essays on World War I: Origins and Prisoners of War* (Brooklyn College Press: New York, 1983), which he coedited with Peter Pastor.

As for the involvement of two of the lesser players in the events of 1914, we now have John F. V. Keiger, *France and the Origins of the First World War* (Macmillan: London, 1983), which concludes that France, "like Britain," entered the war largely for "negative reasons—because not to have done so would have been a certain blow to her national security as it was generally perceived"; and Richard J. B. Bosworth, *Italy and the Approach of the First World War* (St. Martin's Press: New York, 1983), which is particularly good at pointing out how throughout the July crisis, Italy was kept in the dark by its nominal allies, Germany and Austria-Hungary. On the more directly relevant Southeastern European background, some excellent and often very readable material has become available as well. The title that first comes to mind is *Russia's Balkan Entanglements, 1806–1914* (Cambridge University Press: Cambridge and New York, 1991) by Barbara Jelavich, a writer of distinction who has spent a scholarly lifetime in this field. Very much worth consulting among earlier accounts are *Russia and the Balkans: Inter-Balkan Rivalries and Russian Foreign Policy 1908–1914* (University of Toronto Press: Toronto, 1981) by Andrew Rossos, a Canadian scholar; as well as Charles and Barbara Jelavich, *The Establishment of the Balkan National States, 1804–1920* (University of Washington Press: Seattle, 1977); and the classic survey by L. S. Stavrianos, *The Balkans, 1815–1914* (Holt, Rinehart and Winston: New York, 1966).

Among shorter contributions are the present author's somewhat polemical "The Third Balkan War: Origins Reconsidered," in *The Journal of Modern History*, vol. 43, No. 3, September 1971, last reprinted in H. W. Koch, ed., *The Origins of the First World War, Great Power Rivalry and German War Aims*, 2nd ed. (Macmillan: London, 1984); and the response by Paul A. Schroeder, "World War I as Galloping Gertie: A Reply to Joachim Remak," also in *The Journal of Modern History*, vol. 44, No. 3, September 1972. (To identify a quotation here: The passage on p. 134 about how Serbia would "some day set Europe by the ears" comes from the official *British Documents on the Origins of the War*, vol. IX, H. M. Stationery Office: London, 1934, p. 467.)

To see matters from a Habsburg perspective again, two books in particular have stood the test of time very well. They are Arthur J. May, *The Passing of the Hapsburg Monarchy, 1914–1918* (University of Pennsylvania Press: Philadelphia, 1961); and Z. A. B. Zeman, *The Break-up of the Habsburg Empire* (Oxford University Press: London, 1961), both of which, while concentrating on a later period, contain much that is of value for an understanding of the essential problems which the Danubian monarchy faced in 1914.

On a more specialized but crucial topic, it would take extraordinary self-control not to mention Joachim Remak, *Sarajevo, The Story of a Political Murder* (Criterion: New York, 1959), or the same author's interviews with the two then still surviving assassins, "Journey to Sarajevo," *Commentary*, XLVI (July 1968). Vladimir Dedijer, best known for his biography of Tito, has written a book on the assassination and its antecedents, *The Road to Sarajevo* (Simon and Schuster: New York, 1966), which is based on a wide variety of sources and is particularly good in its description of the Bosnian background. While a fair amount of additional material on the assassination has appeared since (again, none of it crucial; what was hidden before has essentially remained hidden) very little of it is available in English, with the exception of David MacKenzie's *Apis: The Congenial Conspirator, The Life of Colonel Dragutin Dimitrijević* (East European Monographs: Boulder, 1989).

Shifting the focus back to Germany, the so-called Fischer controversy still does not seem to have run its course. It was triggered by the Hamburg historian's *Germany's Aims in the First World War* (W. W. Norton: New York, 1967), and its sequel, considered the more important of the two by its author, *World Power or Decline* (Norton: New York, 1974). In both of them, major blame for the war was placed on the

German leadership, for whom, according to Fritz Fischer, it was "now or never" in 1914. The publications that followed, whether pro- or anti-Fischer, clearly are too numerous to list here; perhaps no other work on the topic has caused so much argument. But by way of an introduction both to the thesis itself and to the initial controversy among historians, the strongly pro-Fischer John A. Moses, *The Politics of Illusion, The Fischer Controversy in German Historiography* (Barnes & Noble: New York, 1975) makes a convenient title, while a more balanced account of the topic is provided by Andreas Hillgruber, *Germany and the Two World Wars* (Harvard University Press: Cambridge, MA, 1981), as well as by the selections in H. W. Koch, ed., *The Origins of the First World War, Great Power Rivalry and German War Aims,* mentioned before.

Among the chief witnesses to events at the time in Berlin—a man frequently cited by people on both sides of the Fischer controversy—was the Bethmann aide mentioned in the text. He was Kurt Riezler, later to become a philosopher of note, and an exile from Nazi Germany. His life is well described in Wayne C. Thompson's *In the Eye of the Storm, Kurt Riezler and the Crisis of Modern Germany* (University of Iowa Press: Iowa City, 1980). The book, which quotes widely from Riezler's diaries, with their record of the Chancellor's words, provides a good balance to the Fischer thesis about a war supposedly planned by the German leadership as part of a grand power play. While neither Riezler nor the Chancellor "saw any realistic alternative to supporting Austria," writes Thompson, "Riezler's diary reveals no deliberate attempt on the part of the leading German policymakers to start a war to establish German hegemony in Europe or to clear a path for German territorial annexations."

But the time has come to end this survey by returning to the topic of the war's overall causes again. Several books, both narrative histories and anthologies—all of happily manageable length—will do a fine job of providing that larger picture (and of offering even more suggestions for further reading if those should be wanted). The anthologies are those edited by Holger H. Herwig, H. W. Koch, and Dwight E. Lee mentioned before. The narratives are James Joll, *The Origins of the First World War* (Longman: London, 1984); Laurence Lafore, *The Long Fuse, An Interpretation of the Origins of World War I,* Second Edition (Greenwood Press: Westport, CT, 1981), and L. F. C. Turner, *Origins of the First World War* (Edward Arnold: London, 1970). None of them seem dated; what one notices instead is their spirit of fairness and their sense of style.

By way of a final comment, a note of special appreciation is owed to a number of authors and publishers who have graciously permitted the quotation of copyrighted material cited in this volume. They are Sidney B. Fay, *The Origins of the World War*, The Macmillan Company; Gustav Janouch, *Gespräche mit Kafka*, S. Fischer Verlag; Herman Kahn, *On Thermonuclear War*, Princeton University Press; Theodore Roethke "The Victorians," in *Poetry in Crystal*, Steuben Glass; and A. J. P. Taylor, *The Struggle for Mastery in Europe, 1848–1914*, Oxford University Press; as well as *The Atlantic Monthly*, for the George Kennan quotations, and *Der Monat*, for those by Golo Mann and Johannes Urzidil. It was the latter magazine, too, that prefaced a series of articles on 1914 with a Bismarck sentence that, as far as the copyright laws are concerned, is now in the public domain: "It is not the task of politics to avenge what has happened, but to prevent its recurrence."

# Literary Credits

**vi** Theodore Roethke's poem "The Victorians" is reprinted by permission. (© 1963, Steuben of New York.)

**147** Arthur Guiterman's poem "How I Love to Watch the War Cloud . . ." is reprinted courtesy of the *New York Times*.

# Illustration Credits

**ii** *Simplicissimus*, February 8, 1909.

**2** Kupferstichkabinett der Staatlichen Museen zu Berlin. Preussischer Kulturbesitz. Photo: Jorg P. Anders.

**19** *Punch, or the London Charivari*. March 29, 1890.

**22** *L'Assiette au Beurre*, September, 1911. Cartoon by Jean Veber.

**28** *L'Assiette au Beurre*, July 25, 1908.

**31** *Punch, or the London Charivari*, Oct. 22, 1898.

**39** *Punch, or the London Charivari*, Aug. 2, 1911.

**44** *Simplicissimus*, June 1909. Cartoon by Th. Th. Heine.

**54** *Punch, or the London Charivari*. Oct. 4, 1911.

**56** *Simplicissimus*. Cartoon by Th. Th. Heine.

**62** *Simplicissimus*, 1898. Cartoon by Th. Th. Heine.

**68** *Simplicissimus*, 1904. Cartoon by Ferdinand von Reznicek.

**81** *Simplicissimus*, July 1902. Cartoon by Th. Th. Heine.

**89** *L'Assiette au Beure*, 1908.

**96** Heeresgeschichtliches Museum Militärhistorisches Institut, Vienna.

**98** Austrian National Library.

**124** *Punch, or the London Charivari*, July 29, 1914.

**125** *Punch, or the London Charivari*, Aug. 26, 1914.

**127** *Simplicissimus*.

**130** *Simplicissimus*, April 27, 1915.

**146** *Simplicissimus*, June 1, 1908. Cartoon by Eduard Thöny.

# Index